Environmental Management: Issues and Solutions

Environmental Management: Issues and Solutions

Editor: Timothy Levy

R CALLISTO REFERENCE

www.callistoreference.com

Callisto Reference,
118-35 Queens Blvd., Suite 400,
Forest Hills, NY 11375, USA

Visit us on the World Wide Web at:
www.callistoreference.com

ISBN: 978-1-64116-012-4 (Hardback)

Cataloging-in-Publication Data

Environmental management : issues and solutions / edited by Timothy Levy.
    p. cm.
Includes bibliographical references and index.
ISBN 978-1-64116-012-4
1. Environmental management. 2. Environmental sciences. I. Levy, Timothy.
GE300 .E58 2018
363.705--dc23

Table of Contents

Preface

Pollution has caused irreversible damage to our environment and ecosystem. This calls for immediate action for conservational management of our environment. The sustainable use of natural resources in order to maintain and minimize the damage caused to natural habitats, ecosystem and biodiversity is known as environmental management. This book provides in-depth knowledge about this field. It talks about the different concepts applied in the area. The topics covered in it offer the readers new insights in the field of environmental management. This textbook aims to serve as a resource guide for students and experts alike and contribute to the growth of the discipline.

Given below is the chapter wise description of the book:

Chapter 1- Humans have greatly impacted the biodiversity, natural resources and the Earth as a whole. Because of it, environmental management has become a significant term and its aim is to create a sustainable development plan. The goals of environmental management include solving environment problems such as environmental degradation, protection and upholding the sustainable development of society. This is an introductory section which will introduce briefly all the significant aspects of environmental management.

Chapter 2- Environmental Impact Assessment holds a great impetus in environmental management. It refers to evaluating all the positive and negative aspects of a project or a work on the environment prior to its inauguration. The study is beneficial for project sponsors as it reduces the cost and time for implementation of the project, avoids violations of laws and treatment cost. A healthier environment, decreased resource use, etc are some of the benefits of this procedure. Environmental Impact Assessment is best understood in confluence with the major topics listed in the following section.

Chapter 3- Environmental policies are formed by organizations or governing bodies to oversee work that concerns the environment. Pesticide and insecticide bioaccumulation, overgrazing, food scarcity and climate change are some of the important issues that are taken up in this field. The major policies and laws of environmental management are discussed in this chapter.

Chapter 4- Environmental Management System refers to managing an organization to achieve its environmental goals efficiently. The most widely recognized standards that the system is based on are ISO 14000 and ISO 14001. The chapter also discusses the Life Cycle Assessment (LCA), which is a technique that evaluates the environmental impact a product cause throughout its entire life cycle. The major categories of Environmental Management System Standards and Life Cycle Assessment are dealt with great details in the chapter.

Indeed, my job was extremely crucial and challenging as I had to ensure that every chapter is informative and structured in a student-friendly manner. I am thankful for the support provided by my family and colleagues during the completion of this book.

Editor

Environmental Management: Issues and Goals

Humans have greatly impacted the biodiversity, natural resources and the Earth as a whole. Because of it, environmental management has become a significant term and its aim is to create a sustainable development plan. The goals of environmental management include solving environment problems such as environmental degradation, protection and upholding the sustainable development of society. This is an introductory section which will introduce briefly all the significant aspects of environmental management.

Environmental Issue

Water pollution is an environmental issue that affects many water bodies. This photograph shows foam on the New River as it enters the United States from Mexico.

Environmental issues are harmful effects of human activity on the biophysical environment. Environmental protection is a practice of protecting the natural environment on individual, organizational or governmental levels, for the benefit of both the environment and humans. Environmentalism, a social and environmental movement, addresses environmental issues through advocacy, education and activism.

The carbon dioxide equivalent of greenhouse gases (GHG) in the atmosphere has already exceeded 400 parts per million (NOAA) (with total "long-term" GHG exceeding 455 parts per million). (Intergovernmental Panel on Climate Change Report) This level is considered a tipping point. "The amount of greenhouse gas in the atmosphere is already above the threshold that can potentially cause dangerous climate change. We are already at risk of many areas of pollution...It's not next year or next decade, it's now." Report from the UN Office for the Coordination of Humanitarian Affairs (OCHA):

"Climate disasters are on the rise. Around 70 percent of disasters are now climate related – up from around 50 percent from two decades ago.

These disasters take a heavier human toll and come with a higher price tag. In the last decade, 2.4 billion people were affected by climate related disasters, compared to 1.7 billion in the previous decade. The cost of responding to disasters has risen tenfold between 1992 and 2008. Destructive sudden heavy rains, intense tropical storms, repeated flooding and droughts are likely to increase, as will the vulnerability of local communities in the absence of strong concerted action." "Climate change is not just a distant future threat. It is the main driver behind rising humanitarian needs and we are seeing its impact. The number of people affected and the damages inflicted by extreme weather has been unprecedented."

Environment destruction caused by humans is a global problem, and this is a problem that is on going every day. By year 2050, the global human population is expected to grow by 2 billion people, thereby reaching a level of 9.6 billion people (Living Blue Planet 24). The human effects on Earth can be seen in many different ways. A main one is the temperature rise, and according to the report "Our Changing Climate", the global warming that has been going on for the past 50 years is primarily due to human activities (Walsh, et al. 20). Since 1895, the U.S. average temperature has increased from 1.3 °F to 1.9 °F, with most of the increase taken place since around year 1970 (Walsh, et al. 20).

Types

Major current environmental issues may include climate change, pollution, environmental degradation, and resource depletion etc. The conservation movement lobbies for protection of endangered species and protection of any ecologically valuable natural areas, genetically modified foods and global warming.

Scientific Grounding

The level of understanding of Earth has increased markebly in recent times through science especially with the application of the scientific method. Environmental science is now a multi-disciplinary academic study taught and researched at many universities. This is used as a basis for addressing environmental issues.

Large amounts of data have been gathered and these are collated into reports, of which a common type is the State of the Environment publications. A recent major report was the Millennium Ecosystem Assessment, with input from 1200 scientists and released in 2005, which showed the high level of impact that humans are having on ecosystem services.

Organizations

Environmental issues are addressed at a regional, national or international level by government organizations.

The largest international agency, set up in 1972, is the United Nations Environment Programme. The International Union for Conservation of Nature brings together 83 states, 108 government agencies, 766 Non-governmental organizations and 81 international organizations and about

10,000 experts and scientists from countries around the world. International non-governmental organizations include Greenpeace, Friends of the Earth and World Wide Fund for Nature. Governments enact environmental policy and enforce environmental law and this is done to differing degrees around the world.

Costs

Solutions

> " The only question is whether [the world's environmental problems] will become resolved in pleasant ways of our own choice, or in unpleasant ways not of our choice, such as warfare, genocide, starvation, disease epidemics, and collapses of societies.
>
> "
>
> — Jared Diamond, Collapse: How Societies Choose to Fail or Survive

Sustainability is the key to prevent or reduce the effect of environmental issues. There is now clear scientific evidence that humanity is living unsustainably, and that an unprecedented collective effort is needed to return human use of natural resources to within sustainable limits. For humans to live sustainably, the Earth's natural resources must be used at a rate at which they can be replenished (and by limiting global warming).

Concerns for the environment have prompted the formation of green parties, political parties that seek to address environmental issues. Initially these were formed in Australia, New Zealand and Germany but are now present in many other countries.

Film and Television

There are an increasing number of films being produced on environmental issues, especially on climate change and global warming. Al Gore's 2006 film An Inconvenient Truth gained commercial success and a high media profile.

Environmental Degradation

Eighty-plus years after the abandonment of Wallaroo Mines (Kadina, South Australia), mosses remain the only vegetation at some spots of the site's grounds

Environmental degradation is the deterioration of the environment through depletion of resources such as air, water and soil; the destruction of ecosystems; habitat destruction; the extinction of wildlife; and pollution. It is defined as any change or disturbance to the environment perceived to be deleterious or undesirable. As indicated by the I=PAT equation, environmental impact (I) or degradation is caused by the combination of an already very large and increasing human population (P), continually increasing economic growth or per capita affluence (A), and the application of resource depleting and polluting technology (T).

Environmental degradation is one of the ten threats officially cautioned by the High-level Panel on Threats, Challenges and Change of the United Nations. The United Nations International Strategy for Disaster Reduction defines environmental degradation as "The reduction of the capacity of the environment to meet social and ecological objectives, and needs". Environmental degradation is of many types. When natural habitats are destroyed or natural resources are depleted, the environment is degraded. Efforts to counteract this problem include environmental protection and environmental resources management.

Water Degradation

One major component of environmental degradation is the depletion of the resource of fresh water on Earth. Approximately only 2.5% of all of the water on Earth is fresh water, with the rest being salt water. 69% of the fresh water is frozen in ice caps located on Antarctica and Greenland, so only 30% of the 2.5% of fresh water is available for consumption. Fresh water is an exceptionally important resource, since life on Earth is ultimately dependent on it. Water transports nutrients and chemicals within the biosphere to all forms of life, sustains both plants and animals, and moulds the surface of the Earth with transportation and deposition of materials.

The current top three uses of fresh water account for 95% of its consumption; approximately 85% is used for irrigation of farmland, golf courses, and parks, 6% is used for domestic purposes such as indoor bathing uses and outdoor garden and lawn use, and 4% is used for industrial purposes such as processing, washing, and cooling in manufacturing centers. It is estimated that one in three people over the entire globe are already facing water shortages, almost one-fifth of the world's population live in areas of physical water scarcity, and almost one quarter of the world's population live in a developing country that lacks the necessary infrastructure to use water from available rivers and aquifers. Water scarcity is an increasing problem due to many foreseen issues in the future, including population growth, increased urbanization, higher standards of living, and climate change.

Climate Change and Temperature

Climate change affects the Earth's water supply in a large number of ways. It is predicted that the mean global temperature will rise in the coming years due to a number of forces affecting the climate, the amount of atmospheric CO_2 will rise, and both of these will influence water resources; evaporation depends strongly on temperature and moisture availability, which can ultimately affect the amount of water available to replenish groundwater supplies.

Transpiration from plants can be affected by a rise in atmospheric CO_2, which can decrease their use of water, but can also raise their use of water from possible increases of leaf area. Temperature increase can decrease the length of the snow season in the winter and increase the intensity of snowmelt in warmer seasons, leading to peak runoff of snowmelt earlier in the season, affecting soil moisture, flood and drought risks, and storage capacities depending on the area.

Warmer winter temperatures cause a decrease in snowpack, which can result in diminished water resources during summer. This is especially important at mid-latitudes and in mountain regions that depend on glacial runoff to replenish their river systems and groundwater supplies, making these areas increasingly vulnerable to water shortages over time; an increase in temperature will initially result in a rapid rise in water melting from glaciers in the summer, followed by a retreat in glaciers and a decrease in the melt and consequently the water supply every year as the size of these glaciers get smaller and smaller.

Thermal expansion of water and increased melting of oceanic glaciers from an increase in temperature gives way to a rise in sea level, which can affect the fresh water supply of coastal areas as well; as river mouths and deltas with higher salinity get pushed further inland, an intrusion of saltwater results in an increase of salinity in reservoirs and aquifers. Sea-level rise may also consequently be caused by a depletion of groundwater, as climate change can affect the hydrologic cycle in a number of ways. Uneven distributions of increased temperatures and increased precipitation around the globe results in water surpluses and deficits, but a global decrease in groundwater suggests a rise in sea level, even after meltwater and thermal expansion were accounted for, which can provide a positive feedback to the problems sea-level rise causes to fresh-water supply.

A rise in air temperature results in a rise in water temperature, which is also very significant in water degradation, as the water would become more susceptible to bacterial growth. An increase in water temperature can also affect ecosystems greatly because of a species' sensitivity to temperature, and also by inducing changes in a body of water's self-purification system from decreased amounts of dissolved oxygen in the water due to rises in temperature.

Climate Change and Precipitation

A rise in global temperatures is also predicted to correlate with an increase in global precipitation, but because of increased runoff, floods, increased rates of soil erosion, and mass movement of land, a decline in water quality is probable, while water will carry more nutrients, it will also carry more contaminants. While most of the attention about climate change is directed towards global warming and greenhouse effect, some of the most severe effects of climate change are likely to be from changes in precipitation, evapotranspiration, runoff, and soil moisture. It is generally expected that, on average, global precipitation will increase, with some areas receiving increases and some decreases.

Climate models show that while some regions should expect an increase in precipitation, such as in the tropics and higher latitudes, other areas are expected to see a decrease, such as in the subtropics; this will ultimately cause a latitudinal variation in water distribution. The areas receiving more precipitation are also expected to receive this increase during their winter and actually become drier during their summer, creating even more of a variation of precipitation distribution. Naturally, the distribution of precipitation across the planet is very uneven, causing constant variations in water availability in respective locations.

Changes in precipitation affect the timing and magnitude of floods and droughts, shift runoff processes, and alter groundwater recharge rates. Vegetation patterns and growth rates will be directly affected by shifts in precipitation amount and distribution, which will in turn affect agriculture as well as natural ecosystems. Decreased precipitation will deprive areas of water, causing water tables to fall and reservoirs and wetlands, rivers, and lakes to empty, and possibly an increase in evaporation and evapotranspiration, depending on the accompanied rise in temperature. Groundwater reserves will be depleted, and the remaining water has a greater chance of being of poor quality from saline or contaminants on the land surface.

Population Growth

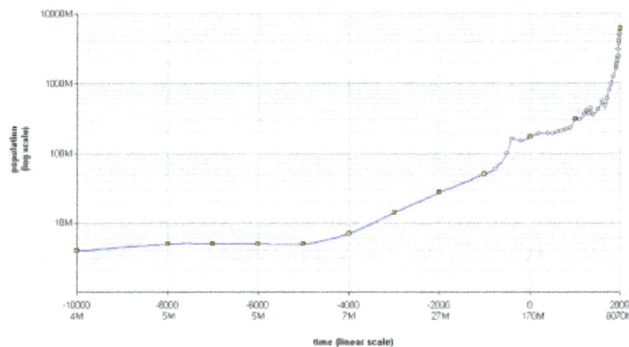

World population growth in a lin-log scale.

The human population on Earth is expanding rapidly which goes hand in hand with the degradation of the environment at large measures. Humanity's appetite for needs is disarranging the environment's natural equilibrium. Production industries are venting smoke and discharging chemicals that are polluting water resources. The smoke that is emitted into the atmosphere holds detrimental gases such as carbon monoxide and sulfur dioxide. The high levels of pollution in the atmosphere form layers that are eventually absorbed into the atmosphere. Organic compounds such as chlorofluorocarbons (CFC's) have generated an unwanted opening in the ozone layer, which emits higher levels of ultraviolet radiation putting the globe at large threat.

The available fresh water being affected by the climate is also being stretched across an ever-increasing global population. It is estimated that almost a quarter of the global population is living in an area that is using more than 20% of their renewable water supply; water use will rise with population while the water supply is also being aggravated by decreases in streamflow and groundwater caused by climate change. Even though some areas may see an increase in freshwater supply from an uneven distribution of precipitation increase, an increased use of water supply is expected.

An increased population means increased withdrawals from the water supply for domestic, agricultural, and industrial uses, the largest of these being agriculture, believed to be the major non-climate driver of environmental change and water deterioration. The next 50 years will likely be the last period of rapid agricultural expansion, but the larger and wealthier population over this time will demand more agriculture.

Population increase over the last two decades, at least in the United States, has also been accompanied by a shift to an increase in urban areas from rural areas, which concentrates the demand for water into certain areas, and puts stress on the fresh water supply from industrial and human contaminants. Urbanization causes overcrowding and increasingly unsanitary living conditions, especially in developing countries, which in turn exposes an increasingly number of people to disease. About 79% of the world's population is in developing countries, which lack access to sanitary water and sewer systems, giving rises to disease and deaths from contaminated water and increased numbers of disease-carrying insects.

Agriculture

Water pollution due to dairy farming in the Wairarapa in New Zealand

Agriculture is dependent on available soil moisture, which is directly affected by climate dynamics, with precipitation being the input in this system and various processes being the output, such as evapotranspiration, surface runoff, drainage, and percolation into groundwater. Changes in climate, especially the changes in precipitation and evapotranspiration predicted by climate models, will directly affect soil moisture, surface runoff, and groundwater recharge.

In areas with decreasing precipitation as predicted by the climate models, soil moisture may be substantially reduced. With this in mind, agriculture in most areas needs irrigation already, which depletes fresh water supplies both by the physical use of the water and the degradation agriculture causes to the water. Irrigation increases salt and nutrient content in areas that would not normally be affected, and damages streams and rivers from damming and removal of water. Fertilizer enters both human and livestock waste streams that eventually enter groundwater, while nitrogen, phosphorus, and other chemicals from fertilizer can acidify both soils and water. Certain agricultural demands may increase more than others with an increasingly wealthier global population, and meat is one commodity expected to double global food demand by 2050, which directly affects the global supply of fresh water. Cows need water to drink, more if the temperature is high and humidity is low, and more if the production system the cow is in is extensive, since finding food takes more effort. Water is needed in processing of the meat, and also in the production of feed for the livestock. Manure can contaminate bodies of freshwater, and slaughterhouses, depending on how well they are managed, contribute waste such as blood, fat, hair, and other bodily contents to supplies of fresh water.

The transfer of water from agricultural to urban and suburban use raises concerns about agricultural sustainability, rural socioeconomic decline, food security, an increased carbon footprint from imported food, and decreased foreign trade balance. The depletion of fresh water, as applied to more specific and populated areas, increases fresh water scarcity among the population and also makes populations susceptible to economic, social, and political conflict in a number of ways; rising sea levels forces migration from coastal areas to other areas farther inland, pushing populations closer together breaching borders and other geographical patterns, and agricultural surpluses and deficits from the availability of water induce trade problems and economies of certain areas. Climate change is an important cause of involuntary migration and forced displacement

Water Management

A stream in the town of Amlwch, Anglesey which is contaminated by acid mine drainage from the former copper mine at nearby Parys Mountain.

The issue of the depletion of fresh water can be met by increased efforts in water management. While water management systems are often flexible, adaptation to new hydrologic conditions may be very costly. Preventative approaches are necessary to avoid high costs of inefficiency and the need for rehabilitation of water supplies, and innovations to decrease overall demand may be important in planning water sustainability.

Water supply systems, as they exist now, were based on the assumptions of the current climate, and built to accommodate existing river flows and flood frequencies. Reservoirs are operated based on past hydrologic records, and irrigation systems on historical temperature, water availability, and crop water requirements; these may not be a reliable guide to the future. Re-examining engineering designs, operations, optimizations, and planning, as well as re-evaluating legal, technical, and economic approaches to manage water resources are very important for the future of water management in response to water degradation. Another approach is water privatization; despite its economic and cultural effects, service quality and overall quality of the water can be more easily controlled and distributed. Rationality and sustainability is appropriate, and requires limits to overexploitation and pollution, and efforts in conservation.

Environmental Protection

Environmental protection is a practice of protecting the natural environment on individual, organisation controlled or governmental levels, for the benefit of both the environment and humans. Due to the pressures of over consumption, population and technology, the biophysical environment is being degraded, sometimes permanently. This has been recognized, and governments have begun placing restraints on activities that cause environmental degradation. Since the 1960s, activity of environmental movements has created awareness of the various environmental issues. There is no agreement on the extent of the environmental impact of human activity and even scientific dishonesty occurs, so protection measures are occasionally debated.

Approaches

Voluntary Environmental Agreements

In industrial countries, voluntary environmental agreements often provide a platform for companies to be recognized for moving beyond the minimum regulatory standards and thus support the development of best environmental practice. For instance, in India, Environment Improvement Trust (EIT) has been working for environment & forest protection since 1998. A group of Green Volunteers get a goal of Green India Clean India concept. CA Gajendra Kumar Jain an Chartered Accountant is founder of Environment Improvement Trust in Sojat city a small village of State of Rajasthan in India. In developing countries, such as throughout Latin America, these agreements are more commonly used to remedy significant levels of non-compliance with mandatory regulation. The challenges that exist with these agreements lie in establishing baseline data, targets, monitoring and reporting. Due to the difficulties inherent in evaluating effectiveness, their use is often questioned and, indeed, the whole environment may well be adversely affected as a result. The key advantage of their use in developing countries is that their use helps to build environmental management capacity.

Ecosystems Approach

An ecosystems approach to resource management and environmental protection aims to consider the complex interrelationships of an entire ecosystem in decision making rather than simply responding to specific issues and challenges. Ideally the decision-making processes under such an approach would be a collaborative approach to planning and decision making that involves a broad range of stakeholders across all relevant governmental departments, as well as representatives of industry, environmental groups and community. This approach ideally supports a better exchange of information, development of conflict-resolution strategies and improved regional conservation.

International Environmental Agreements

Many of the earth's resources are especially vulnerable because they are influenced by human impacts across many countries. As a result of this, many attempts are made by countries to develop agreements that are signed by multiple governments to prevent damage or manage the impacts of human activity on natural resources. This can include agreements that impact factors such as climate, oceans, rivers and air pollution. These international environmental agreements are sometimes legally binding documents that have legal implications when they are not followed and, at other times, are more agreements in principle or are for use as codes of conduct. These agreements have a long history with some multinational agreements being in place from as early as 1910 in Europe, America and Africa. Some of the most well-known international agreements include the Kyoto Protocol and others.

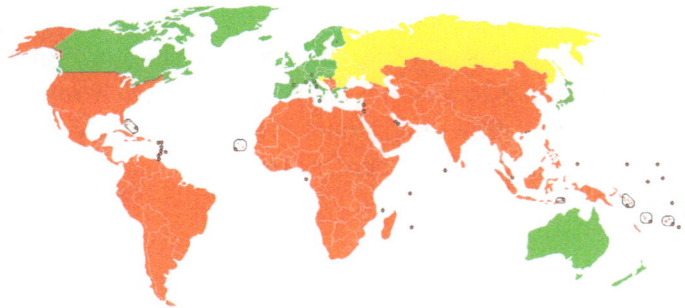

Kyoto Protocol Commitment map 2010

Government

Discussion concerning environmental protection often focuses on the role of government, legislation, and law enforcement. However, in its broadest sense, environmental protection may be seen to be the responsibility of all the people and not simply that of government. Decisions that impact the environment will ideally involve a broad range of stakeholders including industry, indigenous groups, environmental group and community representatives. Gradually, environmental decision-making processes are evolving to reflect this broad base of stakeholders and are becoming more collaborative in many countries.

Many constitutions acknowledge the fundamental right to environmental protection and many international treaties acknowledge the right to live in a healthy environment. Also, many countries have organizations and agencies devoted to environmental protection. There are international environmental protection organizations, such as the United Nations Environment Programme.

Although environmental protection is not simply the responsibility of government agencies, most people view these agencies as being of prime importance in establishing and maintaining basic standards that protect both the environment and the people interacting with it.

Tanzania

Zebras at the Serengeti savana plains in Tanzania

Tanzania is recognised as having some of the greatest biodiversity of any African country. Almost 40% of the land has been established into a network of protected areas, including several national parks. The concerns for the natural environment include damage to ecosystems and loss of habitat resulting from population growth, expansion of subsistence agriculture, pollution, timber extraction and significant use of timber as fuel.

History of Environmental Protection

Environmental protection in Tanzania began during the German occupation of East Africa (1884-1919) — colonial conservation laws for the protection of game and forests were enacted, whereby restrictions were placed upon traditional indigenous activities such as hunting, firewood collecting and cattle grazing. In year 1948, Serengeti was officially established as the first national park for wild cats in East Africa. Since 1983, there has been a more broad-reaching effort to manage environmental issues at a national level, through the establishment of the National Environment Management Council (NEMC) and the development of an environmental act. In 1998 Environment Improvement Trust (EIT) start working for environment & forest protection in India from a small city Sojat. Founder of Environment Improvement Trust is CA Gajendra Kumar Jain working with volunteers.

Government Protection

Division of the biosphere is the main government body that oversees protection. It does this through the formulation of policy, coordinating and monitoring environmental issues, environmental planning and policy-oriented environmental research. The National Environment Management Council (NEMC) is an institution that was initiated when the National Environment Management Act was first introduced in year 1983. This council has the role to advise governments and the international community on a range of environmental issues. The NEMC the following purposes: provide technical advice; coordinate technical activities; develop enforcement guidelines and procedures; assess, monitor and evaluate activities that impact the environment; promote and assist environmental information and communication; and seek advancement of scientific knowledge.

The National Environment Policy of 1997 acts as a framework for environmental decision making in Tanzania. The policy objectives are to achieve the following:

- Ensure sustainable and equitable use of resources without degrading the environment or risking health or safety
- Prevent and control degradation of land, water, vegetation and air
- Conserve and enhance natural and man-made heritage, including biological diversity of unique ecosystems
- Improve condition and productivity of degraded areas
- Raise awareness and understanding of the link between environment and development
- Promote individual and community participation
- Promote international cooperation

Tanzania is a signatory to a significant number of international conventions including the Rio Declaration on Development and Environment 1992 and the Convention on Biological Diversity 1996. The Environmental Management Act, 2004, is the first comprehensive legal and institutional framework to guide environmental-management decisions. The policy tools that are parts of the act includes the use of: environmental-impact assessments, strategics environmentals assessments and taxation on pollution for specific industries and products. The effectiveness of shifing of this act will only become clear over time as concerns regarding its implementation become apparent based on the fact that, historically, there has been a lack of capacity to enforce environmental laws and a lack of working tools to bring environmental-protection objectives into practice.

China

The Longwanqun National Forest Park is a nationally protected nature area in Huinan County, Jilin, China

Formal environmental protection in China House was first stimulated by the 1972 United Nations Conference on the Human Environment held in Stockholm, Sweden. Following this, they began establishing environmental protection agencies and putting controls on some of its industrial waste. China was one of the first developing countries to implement a sustainable development strategy. In 1983 the State Council announced that environmental protection would be one of China's basic national policies and in 1984 the National Environmental Protection Agency (NEPA) was established. Following severe flooding of the Yangtze River basin in 1998, NEPA was upgraded to the State Environmental Protection Agency (SEPA) meaning that environmental protection was now being implemented at a ministerial level. In 2008, SEPA became known by its current name of Ministry of Environmental Protection of the People's Republic of China (MEP).

Pollution control instruments in China

Command-and-control	Economic incentives	Voluntary instruments	Public participation
Concentration-based pollution discharge controls Pollution levy fee Environmental labeling system			Clean-up campaign
Mass-based controls on total provincial discharge	Non-compliance fines	ISO 14000 system	Environmental awareness campaign
Environmental impact assessments (EIA)	Discharge permit system	Cleaner production	Air pollution index

Three synchronization program	Sulfur emission fee	NGOs	Water quality disclosure
Deadline transmission trading		Administrative permission hearing	
Centralized pollution control	Subsidies for energy saving products		
Two compliance policy	Regulation on refuse credit to high-polluting firms		
Environmental compensation fee			

Environmental pollution and ecological degradation has resulted in economic losses for China. In 2005, economic losses (mainly from air pollution) were calculated at 7.7% of China's GDP. This grew to 10.3% by 2002 and the economic loss from water pollution (6.1%) began to exceed that caused by air pollution. China has been one of the top performing countries in terms of GDP growth (9.64% in the past ten years). However, the high economic growth has put immense pressure on its environment and the environmental challenges that China faces are greater than most countries. In 2010 China was ranked 121st out of 163 countries on the Environmental Performance Index.

China has taken initiatives to increase its protection of the environment and combat environmental degradation:

- China's investment in renewable energy grew 18% in 2007 to $15.6 billion, accounting for ~10% of the global investment in this area;

- In 2008, spending on the environment was 1.49% of GDP, up 3.4 times from 2000;

- The discharge of CO (carbon monoxide) and SO2 (sulfur dioxide) decreased by 6.61% and 8.95% in 2008 compared with that in 2005;

- China's protected nature reserves have increased substantially. In 1978 there were only 34 compared with 2,538 in 2010. The protected nature reserve system now occupies 15.5% of the country; this is higher than the world average.

Rapid growth in GDP has been China's main goal during the past three decades with a dominant development model of inefficient resource use and high pollution to achieve high GDP. For China to develop sustainably, environmental protection should be treated as an integral part of its economic policies.

Quote from Shengxian Zhou, head of MEP (2009): "Good economic policy is good environmental policy and the nature of environmental problem is the economic structure, production form and develop model."

European Union

Environmental protection has become an important task for the institutions of the European Community after the Maastricht Treaty for the European Union ratification by all the Member States. The EU is already very active in the field of environmental policy with important directives like those on environmental impact assessment and on the access to environmental information for citizens in the Member States.

Russia

In Russia, environmental protection is considered an integral part of national safety. There is an authorized state body - the Federal Ministry of Natural Resources and Ecology. However, there are a lot of environmental problems.

Latin America

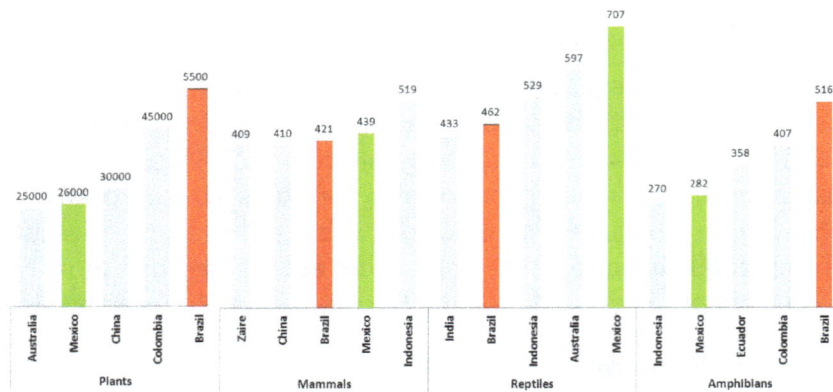

Top 5 Countries by biological diversity

The United Nations Environment Programme (UNEP) has identified 17 megadiverse countries. The list includes six Latin American countries: Brazil, Colombia, Ecuador, Mexico, Peru and Venezuela. Mexico and Brazil stand out among the rest because they have the largest area, population and number of species. These countries represent a major concern for environmental protection because they have high rates of deforestation, ecosystems loss, pollution, and population growth.

Brazil

Panorama of the Iguazu falls in Brazil

Brazil has the largest amount of the world's tropical forests, 4,105,401 km2 (48.1% of Brazil), concentrated in the Amazon region. Brazil is home to vast biological diversity, first among the megadiverse countries of the world, having between 15%-20% of the 1.5 million globally described species.

The organization in charge of environment protection is the Brazilian Ministry of the Environment (in Portuguese: Ministério do Meio Ambiente, MMA). It was first created in year 1973 with the name Special Secretariat for the Environment (Secretaria Especial de Meio Ambiente), changing names several times, and adopting the final name in year 1999. The Ministry is responsible for addressing the following issues:

- A national policy for the environment and for water resources;

- A policy for the preservation, conservation and sustainable use of ecosystems, biodiversity and forests;

- Proposing strategies, mechanisms, economic and social instruments for improving environmental quality, and sustainable use of natural resources;

- Policies for integrating production and the environment;

- Environmental policies and programs for the Legal Amazon;

- Ecological and economic territorial zoning.

In 2011, protected areas of the Amazon covered 2,197,485 km2 (an area larger than Greenland), with conservation units, like national parks, accounting for just over half (50.6%), and indigenous territories representing the remaining 49.4%.

Mexico

The axolotl is an endemic species from the central part of Mexico

With over 200,000 different species, Mexico is home to 10–12% of the world's biodiversity, ranking first in reptile biodiversity and second in mammals—one estimate indicates that over 50% of all animal and plant species live in Mexico.

The history of environmental policy in Mexico started in the 1940s with the enactment of the Law of Conservation of Soil and Water (in Spanish: Ley de Conservación de Suelo y Agua). Three decades later, at the beginning of the 1970s, the Law to Prevent and Control Environmental Pollution was created (Ley para Prevenir y Controlar la Contaminación Ambiental).

In year 1972 was the first direct response from the federal government to address eminent health effects from environmental issues. It established the administrative organization of the Secretariat for the Improvement of the Environment (Subsecretaría para el Mejoramiento del Ambiente) in the Department of Health and Welfare.

The Secretariat of Environment and Natural Resources (Secretaría del Medio Ambiente y Recursos Naturales, SEMARNAT) is Mexico's environment ministry. The Ministry is responsible for addressing the following issues:

- Promote the protection, restoration and conservation of ecosystems, natural resources, goods and environmental services, and to facilitate their use and sustainable development.

- Develop and implement a national policy on natural resources

- Promote environmental management within the national territory, in coordination with all levels of government and the private sector.

- Evaluate and provide determination to the environmental impact statements for development projects and prevention of ecological damage

- Implement national policies on climate change and protection of the ozone layer.

- Direct work and studies on national meteorological, climatological, hydrological, and geo-hydrological systems, and participate in international conventions on these subjects.

- Regulate and monitor the conservation of waterways

In November 2000 there were 127 protected areas; currently there are 174, covering an area of 25,384,818 hectares, increasing federally protected areas from 8.6% to 12.85% its land area.

Oceania

Australia

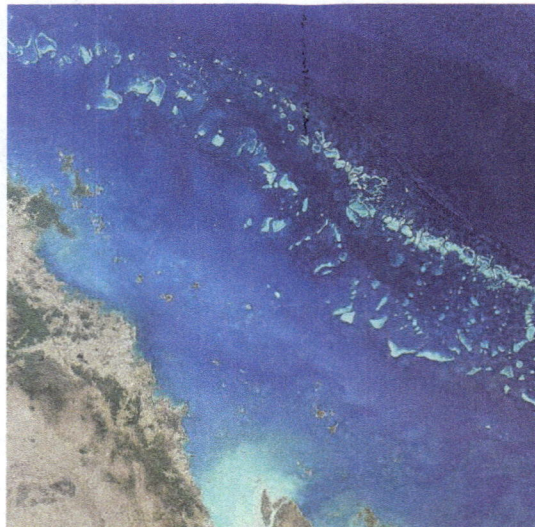

The Great Barrier Reef in Australia is the largest barrier reef in the world

In 2008, there was 98,487,116 ha of terrestrial protected area, covering 12.8% of the land area of Australia. The 2002 figures of 10.1% of terrestrial area and 64,615,554 ha of protected marine area were found to poorly represent about half of Australia's 85 bioregions.

Environmental protection in Australia could be seen as starting with the formation of the first National Park, Royal National Park, in 1879. More progressive environmental protection had it start in the 1960s and 1970s with major international programs such as the United Nations Conference on the Human Environment in 1972, the Environment Committee of the OECD in 1970, and the United Nations Environment Programme of 1972. These events laid the foundations by increasing public awareness and support for regulation. State environmental legislation was irregular and deficient until the Australian Environment Council (AEC) and Council of Nature Conservation Ministers (CONCOM) were established in 1972 and 1974, creating a forum to assist in coordinating environmental and conservation policies between states and neighbouring countries. These councils have since been replaced by the Australian and New Zealand Environment and Conservation Council (ANZECC) in 1991 and finally the Environment Protection and Heritage Council (EPHC) in 2001.

At a national level, the Environment Protection and Biodiversity Conservation Act 1999 is the primary environmental protection legislation for the Commonwealth of Australia. It concerns matters of national and international environmental significance regarding flora, fauna, ecological communities and cultural heritage. It also has jurisdiction over any activity conducted by the Commonwealth, or affecting it, that has significant environmental impact. The act covers eight main areas:

- National Heritage Sites
- World Heritage Sites
- RAMSAR wetlands
- Nationally endangered or threatened species and ecological communities
- Nuclear activities and actions
- Great Barrier Reef Marine Park
- Migratory species
- Commonwealth Marine areas

There are several Commonwealth protected lands due to partnerships with traditional native owners, such as Kakadu National Park, extraordinary biodiversity such as Christmas Island National Park, or managed cooperatively due to cross-state location, such as the Australian Alps National Parks and Reserves.

At a state level, the bulk of environmental protection issues are left to the responsibility of the state or territory. Each state in Australia has its own environmental protection legislation and corresponding agencies. Their jurisdiction is similar and covers point-source pollution, such as from industry or commercial activities, land/water use, and waste management. Most protected lands are managed by states and territories with state legislative acts creating different degrees and definitions of protected areas such as wilderness, national land and marine parks, state forests, and conservation areas. States also create regulation to limit and provide general protection from air, water, and sound pollution.

At a local level, each city or regional council has responsibility over issues not covered by state or national legislation. This includes non-point source, or diffuse pollution, such as sediment pollution from construction sites.

Australia ranks second place on the UN 2010 Human Development Index and one of the lowest debt to GDP ratios of the developed economies. This could be seen as coming at the cost of the environment, with Australia being the world leader in coal exportation and species extinctions. Some have been motivated to proclaim it is Australia's responsibility to set the example of environmental reform for the rest of the world to follow.

New Zealand

At a national level, the Ministry for the Environment is responsible for environmental policy and the Department of Conservation addresses conservation issues. At a regional level the regional councils administer the legislation and address regional environmental issues.

Switzerland

The environmental protection in Switzerland is mainly based on the measures to be taken against global warming. The pollution in Switzerland is mainly the pollution caused by vehicles and the litteration by tourists.

United States

Yosemite National Park in California. One of the first protected areas in the United States

Since 1969, the United States Environmental Protection Agency (EPA) has been working to protect the environment and human health. All U.S. states have their own state departments of environmental protection.

The EPA has drafted "Seven Priorities for EPA's Future", which are:

- "Taking Action on Climate Change"

- "Improving Air Quality"

- "Assuring the Safety of Chemicals"

- "Cleaning Up Our Communities"

- "Protecting America's Waters"

- "Expanding the Conversation on Environmentalism and Working for Environmental Justice"

- "Building Strong State and Tribal Partnerships"

In literature

There are many works of literature that contain the themes of environmental protection but some have been fundamental to its evolution. Several pieces such as A Sand County Almanac by Aldo Leopold, Tragedy of the commons by Garrett Hardin, and Silent Spring by Rachel Carson have become classics due to their far reaching influences. Environmental protection is present in fiction as well as non-fictional literature. Books such as Antarctica and Blockade have environmental protection as subjects whereas The Lorax has become a popular metaphor for environmental protection. "The Limits of Trooghaft" by Desmond Stewart is a short story that provides insight into human attitudes towards animals. Another book called "The Martian Chronicles" by Ray Bradbury investigates issues such as bombs, wars, government control, and what effects these can have on the environment.

Environmental Resource Management

The shrinking Aral Sea, an example of poor water resource management diverted for irrigation.

Environmental resource management is the management of the interaction and impact of human societies on the environment. It is not, as the phrase might suggest, the management of the environment itself. Environmental resources management aims to ensure that ecosystem services are protected and maintained for future human generations, and also maintain ecosystem integrity through considering ethical, economic, and scientific (ecological) variables. Environmental resource management tries to identify factors affected by conflicts that rise between meeting needs and protecting resources. It is thus linked to environmental protection, sustainability and integrated landscape management.

Significance

Environmental resource management is an issue of increasing concern, as reflected in its prevalence in seminal texts influencing global sociopolitical frameworks such as the Brundtland Commission's Our Common Future, which highlighted the integrated nature of environment and international development and the Worldwatch Institute's annual State of the World reports.

The environment determines the nature of people, animals, plants, and places around the Earth, affecting behaviour, religion, culture and economic practices.

Scope

Improved agricultural practices such as these terraces in northwest Iowa can serve to preserve soil and improve water quality

Environmental resource management can be viewed from a variety of perspectives. It involves the management of all components of the biophysical environment, both living (biotic) and non-living (abiotic), and the relationships among all living species and their habitats. The environment also involves the relationships of the human environment, such as the social, cultural and economic environment, with the biophysical environment. The essential aspects of environmental resource management are ethical, economical, social, and technological. These underlie principles and help make decisions.

The concept of environmental determinism, probabilism and possibilism are significant in the concept of environmental resource management.

Environmental resource management covers many areas in science, including geography, biology, physics, chemistry, sociology, psychology, and physiology.

Aspects

Ethical

Environmental resource management strategies are intrinsically driven by conceptions of human-nature relationships. Ethical aspects involve the cultural and social issues relating to the environment, and dealing with changes to it. "All human activities take place in the context of certain types of relationships between society and the bio-physical world (the rest of nature)," and so, there is a great significance in understanding the ethical values of different groups around the world. Broadly speaking, two schools of thought exist in environmental ethics: Anthropocentrism and Ecocentrism, each influencing a broad spectrum of environmental resource management styles along a continuum. These styles perceive "...different evidence, imperatives, and problems, and prescribe different solutions, strategies, technologies, roles for economic sectors, culture, governments, and ethics, etc."

Anthropocentrism

Anthropocentrism, "...an inclination to evaluate reality exclusively in terms of human values," is an ethic reflected in the major interpretations of Western religions and the dominant economic paradigms of the industrialised world. Anthropocentrism looks at nature as existing solely for the benefit of humans, and as a commodity to use for the good of humanity and to improve human quality of life. Anthropocentric environmental resource management is therefore not the conservation of the environment solely for the environment's sake, but rather the conservation of the environment, and ecosystem structure, for humans' sake.

Ecocentrism

Ecocentrists believe in the intrinsic value of nature while maintaining that human beings must use and even exploit nature to survive and live. It is this fine ethical line that ecocentrists navigate between fair use and abuse. At an extreme of the ethical scale, ecocentrism includes philosophies such as ecofeminism and deep ecology, which evolved as a reaction to dominant anthropocentric paradigms. "In its current form, it is an attempt to synthesize many old and some new philosophical attitudes about the relationship between nature and human activity, with particular emphasis on ethical, social, and spiritual aspects that have been downplayed in the dominant economic worldview."

Economics

A water harvesting system collects rainwater from the Rock of Gibraltar into pipes that lead to tanks excavated inside the rock.

The economy functions within, and is dependent upon goods and services provided by natural ecosystems. The role of the environment is recognized in both classical economics and neoclassical economics theories, yet the environment was a lower priority in economic policies from 1950 to 1980 due to emphasis from policy makers on economic growth. With the prevalence of environmental problems, many economists embraced the notion that, "If environmental sustainability must coexist for economic sustainability, then the overall system must [permit] identification of an equilibrium between the environment and the economy." As such, economic policy makers began to incorporate the functions of the natural environment—or natural capital — particularly as a sink for wastes and for the provision of raw materials and amenities.

Debate continues among economists as to how to account for natural capital, specifically whether resources can be replaced through knowledge and technology, or whether the environment is a closed system that cannot be replenished and is finite. Economic models influence environmental resource management, in that management policies reflect beliefs about natural capital scarcity. For someone who believes natural capital is infinite and easily substituted, environmental management is irrelevant to the economy. For example, economic paradigms based on neoclassical models of closed economic systems are primarily concerned with resource scarcity, and thus prescribe legalizing the environment as an economic externality for an environmental resource management strategy. This approach has often been termed 'Command-and-control'. Colby has identified trends in the development of economic paradigms, among them, a shift towards more ecological economics since the 1990s.

Ecology

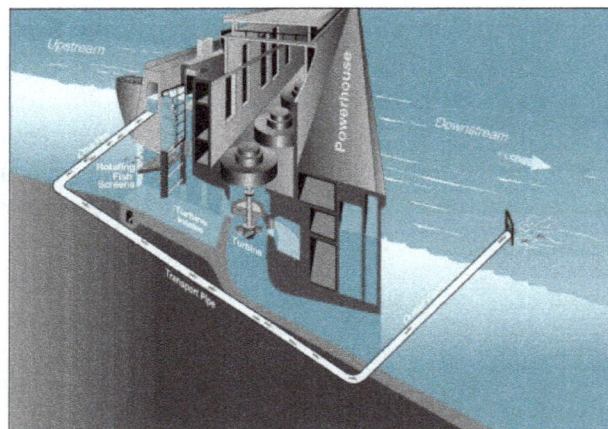

A diagram showing the juvenile fish bypass system, which allows young salmon and steelhead to safely pass the Rocky Reach Hydro Project in Washington

"The pairing of significant uncertainty about the behaviour and response of ecological systems with urgent calls for near-term action constitutes a difficult reality, and a common lament" for many environmental resource managers. Scientific analysis of the environment deals with several dimensions of ecological uncertainty. These include: structural uncertainty resulting from the misidentification, or lack of information pertaining to the relationships between ecological variables; parameter uncertainty referring to "uncertainty associated with parameter values that are not known precisely but can be assessed and reported in terms of the likelihood...of experiencing a defined range of outcomes"; and stochastic uncertainty stemming from chance or unrelated factors. Adaptive management is considered a useful framework for dealing with situations of high levels of uncertainty though it is not without its detractors.

Fencing separates big game from vehicles along the Quebec Autoroute 73 in Canada.

A common scientific concept and impetus behind environmental resource management is carrying capacity. Simply put, carrying capacity refers to the maximum number of organisms a particular resource can sustain. The concept of carrying capacity, whilst understood by many cultures over history, has its roots in Malthusian theory. An example is visible in the EU Water Framework Directive. However, it is argued that Western scientific knowledge ... is often insufficient to deal with the full complexity of the interplay of variables in environmental resource management. These concerns have been recently addressed by a shift in environmental resource management approaches to incorporate different knowledge systems including traditional knowledge, reflected in approaches such as adaptive co-management community-based natural resource management and transitions management. among others.

Sustainability

Sustainability in environmental resource management involves managing economic, social, and ecological systems both within and outside an organizational entity so it can sustain itself and the system it exists in. In context, sustainability implies that rather than competing for endless growth on a finite planet, development improves quality of life without necessarily consuming more resources. Sustainably managing environmental resources requires organizational change that instills sustainability values that portrays these values outwardly from all levels and reinforces them to surrounding stakeholders. The end result should be a symbiotic relationship between the sustaining organization, community, and environment.

Many drivers compel environmental resource management to take sustainability issues into account. Today's economic paradigms do not protect the natural environment, yet they deepen human dependency on biodiversity and ecosystem services. Ecologically, massive environmental degradation and climate change threaten the stability of ecological systems that humanity depends on. Socially, an increasing gap between rich and poor and the global North-South divide denies many access to basic human needs, rights, and education, leading to further environmental destruction. The planet's unstable condition is caused by many anthropogenic sources. As an exceptionally powerful contributing factor to social and environmental change, the modern organisation has the potential to apply environmental resource management with sustainability principals to achieve highly effective outcomes. To achieve sustainable development with environmental resource management an organisation should work within sustainability principles, including social and environmental accountability, long-term planning; a strong, shared vision; a holistic focus; devolved and consensus decision making; broad stakeholder engagement and justice; transparency measures; trust; and flexibility.

Current Paradigm Shifts

To adjust to today's environment of quick social and ecological changes, some organizations have begun to experiment with new tools and concepts. Those that are more traditional and stick to hierarchical decision making have difficulty dealing with the demand for lateral decision making that supports effective participation. Whether it be a matter of ethics or just strategic advantage organizations are internalizing sustainability principles. Some of the world's largest and most profitable corporations are shifting to sustainable environmental resource management: Ford, Toyota, BMW, Honda, Shell, Du Pont, Statoil, Swiss Re, Hewlett-Packard, and Unilever, among others. An extensive study by the Boston Consulting Group reaching 1,560 business leaders from diverse regions, job positions, expertise in sustainability, industries, and sizes of organizations, revealed the many benefits of sustainable practice as well as its viability.

It is important to note that though sustainability of environmental resource management has improved, corporate sustainability, for one, has yet to reach the majority of global companies operating in the markets. The three major barriers to preventing organizations to shift towards sustainable practice with environmental resource management are not understanding what sustainability is; having difficulty modeling an economically viable case for the switch; and having a flawed execution plan, or a lack thereof. Therefore, the most important part of shifting an organization to adopt sustainability in environmental resource management would be to create a shared vision and understanding of what sustainability is for that particular organization, and to clarify the business case.

Stakeholders

Public Sector

A conservation project in North Carolina involving the search for bog turtles was conducted by United States Fish and Wildlife Service and the North Carolina Wildlife Resources Commission and its volunteers

The public sector comprises the general government sector plus all public corporations including the central bank. In environmental resource management the public sector is responsible for administering natural resource management and implementing environmental protection legislation. The traditional role of the public sector in environmental resource management is to provide professional judgement through skilled technicians on behalf of the public. With the increase of intractable environmental problems, the public sector has been led to examine alternative paradigms for managing environmental resources. This has resulted in the public sector working collaboratively with other sectors (including other governments, private and civil) to encourage sustainable natural resource management behaviours.

Private Sector

The private sector comprises private corporations and non-profit institutions serving households. The private sector's traditional role in environmental resource management is that of the recovery of natural resources. Such private sector recovery groups include mining (minerals and petroleum), forestry and fishery organisations. Environmental resource management undertaken by the private sectors varies dependent upon the resource type, that being renewable or non-renewable and private and common resources. Environmental managers from the private sector also need skills to manage collaboration within a dynamic social and political environment.

Civil Society

Civil society comprises associations in which societies voluntarily organise themselves into and which represent a wide range of interests and ties. These can include community-based organisations, indigenous peoples' organisations and non-government organisations (NGO). Functioning through strong public pressure, civil society can exercise their legal rights against the implementation of resource management plans, particularly land management plans. The aim of civil society in environmental resource management is to be included in the decision-making process by means of public participation. Public participation can be an effective strategy to invoke a sense of social responsibility of natural resources.

Tools

As with all management functions, effective management tools, standards and systems are required. An environmental management standard or system or protocol attempts to reduce environmental impact as measured by some objective criteria. The ISO 14001 standard is the most widely used standard for environmental risk management and is closely aligned to the European Eco-Management and Audit Scheme (EMAS). As a common auditing standard, the ISO 19011 standard explains how to combine this with quality management.

Other environmental management systems (EMS) tend to be based on the ISO 14001 standard and many extend it in various ways:

- The Green Dragon Environmental Management Standard is a five-level EMS designed for smaller organisations for whom ISO 14001 may be too onerous and for larger organisations who wish to implement ISO 14001 in a more manageable step-by-step approach,

- BS 8555 is a phased standard that can help smaller companies move to ISO 14001 in six manageable steps,

- The Natural Step focuses on basic sustainability criteria and helps focus engineering on reducing use of materials or energy use that is unsustainable in the long term,

- Natural Capitalism advises using accounting reform and a general biomimicry and industrial ecology approach to do the same thing,

- US Environmental Protection Agency has many further terms and standards that it defines as appropriate to large-scale EMS,

- The UN and World Bank has encouraged adopting a "natural capital" measurement and management framework,

- The European Union Eco-Management and Audit Scheme (EMAS).

Other strategies exist that rely on making simple distinctions rather than building top-down management "systems" using performance audits and full cost accounting. For instance, Ecological Intelligent Design divides products into consumables, service products or durables and unsaleables — toxic products that no one should buy, or in many cases, do not realize they are buying. By eliminating the unsaleables from the comprehensive outcome of any purchase, better environmental resource management is achieved without systems.

Recent successful cases have put forward the notion of integrated management. It shares a wider approach and stresses out the importance of interdisciplinary assessment. It is an interesting notion that might not be adaptable to all cases.

Environment Management (EM)

The natural environment, harbouring myriad life forms and ecosystems, is an outcome of millions of years of evolution and mutation. We depend on the environment to meet our basic requirements such as food, fuel, fibre, fodder, minerals and vital support systems (e.g., water and air). However, due to our continuous indulgence in the exploitation of nature for immediate gains, we contribute to environmental degradation and depletion of natural resources, and this should be arrested. It is, therefore,

essential that we understand the function and interaction of physical and biological elements of the environment and apply this knowledge for sound management programmes to conserve the natural resources we are endowed with. Sustainable management of renewable resources in industry, agriculture, recreation, health, forestry, fisheries, education or urban development is, therefore, critical. The conservation of resources and maintenance of the quality of ecosystem requires as much imaginative and innovative technology as the active participation of the public.

Environment management thus involves managing the environment while ensuring the prudent use of natural resources without reducing their productivity and quality. Due to impacts of modern human society the environment is depreciating in life sustaining capabilities at ever increasing pace, and this should be regulated. It is therefore, essential to understand the function and interaction of physical and biological elements of the environment and develop management programs to conserve natural resources. Objective of this course is to explore ways to reduce environmental footprints in day to day activities. Sustainable environmental management of renewable resources oceans, atmosphere, freshwater, land, energy, food, materials, forestry, fisheries, etc.. Major requirements to attain sustainability are to develop innovative and imaginative technologies in an ecofriendly system of working along with active participation of the public towards the environment.

Essentially, EM represents the management of various activities, including environmental action plan, conservation of resources, environmental status evaluation and environmental legislation and administration, and focuses more on implementation, monitoring, auditing, and practice and real-world issues than on theoretical planning.

Definition and Scope

Different people define environmental management differently, depending on the context in which they work. For the purposes of this Course, let us consider the following two views:

(i) The process of allocating natural and artificial resources so as to make optimum use of the environment in satisfying basic human needs at the minimum, and more if possible, on a sustainable basis (Jolly, 1978).

(ii) A generic description of a process undertaken by systemsoriented professionals with a natural science, social science, or less commonly, an engineering, law, or design background, tackling problems of the human altered environment on an interdisciplinary basis from a quantitative and/or futuristic viewpoint (Dorney, 1989).

Some of the characteristics of EM include the following:

- It is often used as a generic term.

- It supports sustainable development.

- It deals with a world affected by human beings.

- It demands a multidisciplinary or interdisciplinary approach.

- It has to integrate different development viewpoints.

- It seeks to integrate science, social science, policy-making and planning.

- It recognizes the desirability of meeting, and if possible exceeding basic human needs.

- The time-scale involved extends beyond the short-term and concerns range from local to global.

- It should show opportunities as well as address threats and problems.

- It stresses stewardship, rather than exploitation.

Role of environmental manager is to interface with ecology, economics, law, politics, etc. while developing EM action plan (key elements include environmental policy, audit, EM systems standards and external standards). In fact planning activities have long term impacts on the environ-ment hence they should be managed at all levels: regional, national and international. This implies that an environmental manager must „think globally and act locally" and adopt a long-term holis-tic outlook.

Goals of EM

The goals of EM, which is an approach for environmental stewardship integrating ecology, poli-cy-making, planning and social development, include:

- preventing and resolving environmental problems;

- establishing limits;

- establishing and nurturing institutions that effectively support environmental research, monitoring and management;

- identifying threats and opportunities;

- sustaining and, if possible, improving existing resources; improving the quality of life;

- Identifying environmentally sound technologies or policies.

Thus, EM demands scoping, i.e., deciding goals and setting limits on efforts. Having touched upon the definition and characteristics of EM, we will discuss some of the crucial issues relating to EM such as its need as a result of drastic population growth, sustainable development and EM tools.

Need for Environment Management

The world population is estimated to be over 6 billion. The growth of human population is continu-ously increasing. In the last 10, 000 years, the population of the world has increased over a 1000-fold and much of that change has occurred in the last century. This phenomenal growth in population has put pressure on the means of subsistence, throwing it out of balance with the environment. The interaction between population and environment is very complex and dynamic. Admittedly, howev-er, we are yet to understand this complex interrelationship at its micro-level and its spatio-temporal ramifications over a region or the globe in its entirety. What is, nevertheless, certain is that as the population grows, the level of consumption of natural resources and production of wastes propor-tionately increases (Harrison, 1990). And, we know that the environment is constantly changing, due to human activities, leading to such problems as soil erosion, floods, droughts, climate changes, desertification and general degradation of the environment. In a finite world, there are limits, and there indeed are complex environment-population linkages and feedback.

Sustainable development is, therefore, necessary to sustain the quality of life without exceeding environmental limits.

Sustainable Development

Sustainable development is the organizing principle for meeting human development goals while at the same time sustaining the ability of natural systems to provide the natural resources and ecosystem services upon which the economy and society depends. The desirable end result is a state of society where living conditions and resource use continue to meet human needs without undermining the integrity and stability of the natural systems.

Wind powers 5 MW wind turbines on a wind farm 28 km off the coast of Belgium.

While the modern concept of sustainable development is derived mostly from the 1987 Brundtland Report, it is also rooted in earlier ideas about sustainable forest management and twentieth century environmental concerns. As the concept developed, it has shifted to focus more on economic development, social development and environmental protection for future generations. It has been suggested that "the term 'sustainability' should be viewed as humanity's target goal of human-ecosystem equilibrium (homeostasis), while 'sustainable development' refers to the holistic approach and temporal processes that lead us to the end point of sustainability."

The concept of sustainable development has been — and still is — subject to criticism. What, exactly, is to be sustained in sustainable development? It has been argued that there is no such thing as a sustainable use of a non-renewable resource, since any positive rate of exploitation will eventually lead to the exhaustion of Earth's finite stock; this perspective renders the Industrial Revolution as a whole unsustainable. It has also been argued that the meaning of the concept has opportunistically been stretched from 'conservation management' to 'economic development', and that the Brundtland Report promoted nothing but a business as usual strategy for world development, with an ambiguous and insubstantial concept attached as a public relations slogan.

History

Sustainability can be defined as the practice of maintaining processes of productivity indefinitely—natural or human made—by replacing resources used with resources of equal or greater value without degrading or endangering natural biotic systems. Sustainable development ties together concern for the carrying capacity of natural systems with the social, political, and economic challenges faced by humanity. Sustainability science is the study of the concepts of sustainable development and environmental science. There is an additional focus on the present generations' responsibility to regenerate, maintain and improve planetary resources for use by future generations.

The Blue Marble photograph, taken from Apollo 17 on 7 December 1972, quickly became an icon of environmental conservation.

Sustainable development has its roots in ideas about sustainable forest management which were developed in Europe during the seventeenth and eighteenth centuries. In response to a growing awareness of the depletion of timber resources in England, John Evelyn argued that "sowing and planting of trees had to be regarded as a national duty of every landowner, in order to stop the destructive over-exploitation of natural resources" in his 1662 essay Sylva. In 1713 Hans Carl von Carlowitz, a senior mining administrator in the service of Elector Frederick Augustus I of Saxony published Sylvicultura oeconomica, a 400-page work on forestry. Building upon the ideas of Evelyn and French minister Jean-Baptiste Colbert, von Carlowitz developed the concept of managing forests for sustained yield. His work influenced others, including Alexander von Humboldt and Georg Ludwig Hartig, eventually leading to the development of a science of forestry. This in turn influenced people like Gifford Pinchot, first head of the US Forest Service, whose approach to forest management was driven by the idea of wise use of resources, and Aldo Leopold whose land ethic was influential in the development of the environmental movement in the 1960s.

Following the publication of Rachel Carson's Silent Spring in 1962, the developing environmental movement drew attention to the relationship between economic growth and development and environmental degradation. Kenneth E. Boulding in his influential 1966 essay The Economics of the Coming Spaceship Earth identified the need for the economic system to fit itself to the ecological system with its limited pools of resources. One of the first uses of the term sustainable in the contemporary sense was by the Club of Rome in 1972 in its classic report on the Limits to Growth, written by a group of scientists led by Dennis and Donella Meadows of the Massachusetts Institute of Technology. Describing the desirable "state of global equilibrium", the authors wrote: "We are searching for a model output that represents a world system that is sustainable without sudden and uncontrolled collapse and capable of satisfying the basic material requirements of all of its people."

In 1980 the International Union for the Conservation of Nature published a world conservation strategy that included one of the first references to sustainable development as a global priority and introduced the term "sustainable development". Two years later, the United Nations World Charter for Nature raised five principles of conservation by which human conduct affecting nature is to be guided and judged. In 1987 the United Nations World Commission on Environment and Development released the report Our Common Future, commonly called the Brundtland Report. The report included what is now one of the most widely recognised definitions of sustainable development.

" Sustainable development is development that meets the needs of the present without compromising the ability of future generations to meet their own needs. It contains within it two key concepts:

The concept of 'needs', in particular, the essential needs of the world's poor, to which overriding priority should be given; and

The idea of limitations imposed by the state of technology and social organization on the environment's ability to meet present and future needs. "

— World Commission on Environment and Development, Our Common Future (1987)

Since the Brundtland Report, the concept of sustainable development has developed beyond the initial intergenerational framework to focus more on the goal of "socially inclusive and environmentally sustainable economic growth". In 1992, the UN Conference on Environment and Development published the Earth Charter, which outlines the building of a just, sustainable, and peaceful global society in the 21st century. The action plan Agenda 21 for sustainable development identified information, integration, and participation as key building blocks to help countries achieve development that recognises these interdependent pillars. It emphasises that in sustainable development everyone is a user and provider of information. It stresses the need to change from old sector-centered ways of doing business to new approaches that involve cross-sectoral co-ordination and the integration of environmental and social concerns into all development processes. Furthermore, Agenda 21 emphasises that broad public participation in decision making is a fundamental prerequisite for achieving sustainable development.

Under the principles of the United Nations Charter the Millennium Declaration identified principles and treaties on sustainable development, including economic development, social development and environmental protection. Broadly defined, sustainable development is a systems approach to growth and development and to manage natural, produced, and social capital for the welfare of their own and future generations. The term sustainable development as used by the United Nations incorporates both issues associated with land development and broader issues of human development such as education, public health, and standard of living.

A 2013 study concluded that sustainability reporting should be reframed through the lens of four interconnected domains: ecology, economics, politics and culture.

The Sustainable Development Goals (SDGs)

In September 2015, the United Nations General Assembly formally adopted the "universal, integrated and transformative" 2030 Agenda for Sustainable Development, a set of 17 Sustainable Development Goals (SDGs). The goals are to be implemented and achieved in every country from the year 2016 to 2030.

Dimensions

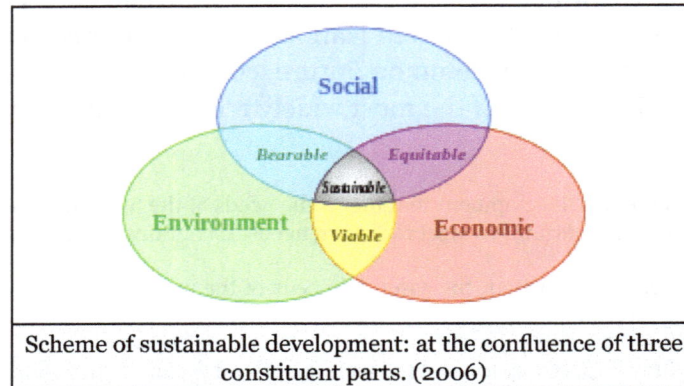

Scheme of sustainable development: at the confluence of three constituent parts. (2006)

Sustainable development, or sustainability, has been described in terms of three spheres, dimensions, domains or pillars, i.e. the environment, the economy and society. The three-sphere framework was initially proposed by the economist René Passet in 1979. It has also been worded as "economic, environmental and social" or "ecology, economy and equity." This has been expanded by some authors to include a fourth pillar of culture, institutions or governance.

Environmental

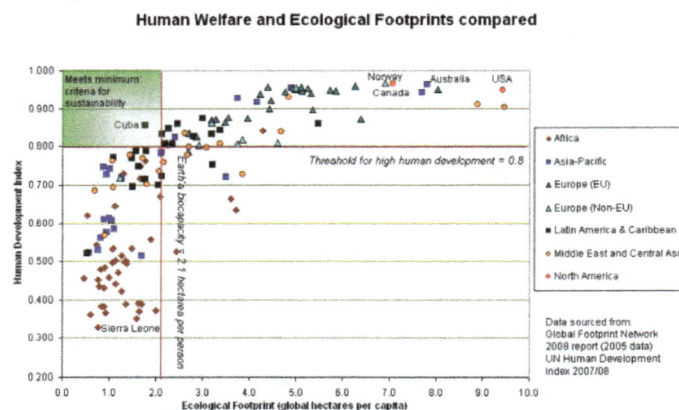

Relationship between ecological footprint and Human Development Index (HDI)

The ecological stability of human settlements is part of the relationship between humans and their natural, social and built environments. Also termed human ecology, this broadens the focus of sustainable development to include the domain of human health. Fundamental human needs such as the availability and quality of air, water, food and shelter are also the ecological foundations for sustainable development; addressing public health risk through investments in ecosystem services can be a powerful and transformative force for sustainable development which, in this sense, extends to all species.

Environmental sustainability concerns the natural environment and how it endures and remains diverse and productive. Since natural resources are derived from the environment, the state of air, water, and the climate are of particular concern. The IPCC Fifth Assessment Report outlines current knowledge about scientific, technical and socio-economic information concerning climate change, and lists options for adaptation and mitigation. Environmental sustainability requires society to design activities to meet human needs while preserving the life support systems of the planet. This, for example, entails using water sustainably, utilizing renewable energy, and sustainable material supplies (e.g. harvesting wood from forests at a rate that maintains the biomass and biodiversity).

An unsustainable situation occurs when natural capital (the sum total of nature's resources) is used up faster than it can be replenished. Sustainability requires that human activity only uses nature's resources at a rate at which they can be replenished naturally. Inherently the concept of sustainable development is intertwined with the concept of carrying capacity. Theoretically, the long-term result of environmental degradation is the inability to sustain human life. Such degradation on a global scale should imply an increase in human death rate until population falls to what the degraded environment can support. If the degradation continues beyond a certain tipping point or critical threshold it would lead to eventual extinction for humanity.

Consumption of non-renewable resources	State of environment	Sustainability
More than nature's ability to replenish	Environmental degradation	Not sustainable
Equal to nature's ability to replenish	Environmental equilibrium	Steady state economy
Less than nature's ability to replenish	Environmental renewal	Environmentally sustainable

Integral elements for a sustainable development are research and innovation activities. A telling example is the European environmental research and innovation policy, which aims at defining and implementing a transformative agenda to greening the economy and the society as a whole so to achieve a truly sustainable development. Research and innovation in Europe is financially supported by the programme Horizon 2020, which is also open to participation worldwide. A promising direction towards sustainable development is to design systems that are flexible and reversible.

Pollution of the public resources is really not a different action, it just is a reverse tragedy of the commons, in that instead of taking something out, something is put into the commons. When the costs of polluting the commons are not calculated into the cost of the items consumed, then it becomes only natural to pollute, as the cost of pollution is external to the cost of the goods produced and the cost of cleaning the waste before it is discharged exceeds the cost of releasing the waste directly into the commons. So, the only way to solve this problem is by protecting the ecology of the commons by making it, through taxes or fines, more costly to release the waste directly into the commons than would be the cost of cleaning the waste before discharge.

So, one can try to appeal to the ethics of the situation by doing the right thing as an individual, but in the absence of any direct consequences, the individual will tend to do what is best for the person and not what is best for the common good of the public. Once again, this issue needs to be addressed. Because, left unaddressed, the development of the commonly owned property will become impossible to achieve in a sustainable way. So, this topic is central to the understanding of creating a sustainable situation from the management of the public resources that are used for personal use.

Agriculture

Sustainable agriculture consists of environment friendly methods of farming that allow the production of crops or livestock without damage to human or natural systems. It involves preventing adverse effects to soil, water, biodiversity, surrounding or downstream resources—as well as to those working or living on the farm or in neighboring areas. The concept of sustainable agriculture extends intergenerationally, passing on a conserved or improved natural resource, biotic, and economic base rather than one which has been depleted or polluted. Elements of sustainable agriculture include permaculture, agroforestry, mixed farming, multiple cropping, and crop rotation.

Numerous sustainability standards and certification systems exist, including organic certification, Rainforest Alliance, Fair Trade, UTZ Certified, Bird Friendly, and the Common Code for the Coffee Community (4C).

Economics

A sewage treatment plant that uses solar energy, located at Santuari de Lluc monastery, Majorca.

It has been suggested that because of rural poverty and overexploitation, environmental resources should be treated as important economic assets, called natural capital. Economic development has traditionally required a growth in the gross domestic product. This model of unlimited personal and GDP growth may be over. Sustainable development may involve improvements in the quality of life for many but may necessitate a decrease in resource consumption. According to ecological economist Malte Faber, ecological economics is defined by its focus on nature, justice, and time. Issues of intergenerational equity, irreversibility of environmental change, uncertainty of long-term outcomes, and sustainable development guide ecological economic analysis and valuation.

As early as the 1970s, the concept of sustainability was used to describe an economy "in equilibrium with basic ecological support systems." Scientists in many fields have highlighted The Limits to Growth, and economists have presented alternatives, for example a 'steady-state economy'; to address concerns over the impacts of expanding human development on the planet. In 1987 the economist Edward Barbier published the study The Concept of Sustainable Economic Development, where he recognised that goals of environmental conservation and economic development are not conflicting and can be reinforcing each other.

A World Bank study from 1999 concluded that based on the theory of genuine savings, policy-makers have many possible interventions to increase sustainability, in macroeconomics or purely environmental. A study from 2001 noted that efficient policies for renewable energy and pollution are compatible with increasing human welfare, eventually reaching a golden-rule steady state. The study, Interpreting Sustainability in Economic Terms, found three pillars of sustainable development, interlinkage, intergenerational equity, and dynamic efficiency.

But Gilbert Rist points out that the World Bank has twisted the notion of sustainable development to prove that economic development need not be deterred in the interest of preserving the ecosystem. He writes: "From this angle, 'sustainable development' looks like a cover-up operation. ... The thing that is meant to be sustained is really 'development', not the tolerance capacity of the ecosystem or of human societies."

The World Bank, a leading producer of environmental knowledge, continues to advocate the win-win prospects for economic growth and ecological stability even as its economists express their doubts. Herman Daly, an economist for the Bank from 1988 to 1994, writes:

When authors of WDR '92 [the highly influential 1992 World Development Report that featured the environment] were drafting the report, they called me asking for examples of "win-win" strategies in my work. What could I say? None exists in that pure form; there are trade-offs, not "win-wins." But they want to see a world of "win-wins" based on articles of faith, not fact. I wanted to contribute because WDRs are important in the Bank, [because] task managers read [them] to find philosophical justification for their latest round of projects. But they did not want to hear about how things really are, or what I find in my work..."

A meta review in 2002 looked at environmental and economic valuations and found a lack of "sustainability policies". A study in 2004 asked if we consume too much. A study concluded in 2007 that knowledge, manufactured and human capital (health and education) has not compensated for the degradation of natural capital in many parts of the world. It has been suggested that intergenerational equity can be incorporated into a sustainable development and decision making, as has become common in economic valuations of climate economics. A meta review in 2009 identified conditions for a strong case to act on climate change, and called for more work to fully account of the relevant economics and how it affects human welfare. According to free-market environmentalist John Baden "the improvement of environment quality depends on the market economy and the existence of legitimate and protected property rights." They enable the effective practice of personal responsibility and the development of mechanisms to protect the environment. The State can in this context "create conditions which encourage the people to save the environment."

Misum, Mistra Center for Sustainable Markets, based at Stockholm School of Economics, aims to provide policy research and advice to Swedish and international actors on Sustainable Markets. Misum is a cross-disciplinary and multi-stakeholder knowledge center dedicated to sustainability and sustainable markets and contains three research platforms: Sustainability in Financial Markets (Mistra Financial Systems), Sustainability in Production and Consumption and Sustainable Socio-Economic Development.

Environmental Economics

The total environment includes not just the biosphere of earth, air, and water, but also human interactions with these things, with nature, and what humans have created as their surroundings.

As countries around the world continue to advance economically, they put a strain on the ability of the natural environment to absorb the high level of pollutants that are created as a part of this economic growth. Therefore, solutions need to be found so that the economies of the world can continue to grow, but not at the expense of the public good. In the world of economics the amount of environmental quality must be considered as limited in supply and therefore is treated as a scarce resource. This is a resource to be protected. One common way to analyze possible outcomes of policy decisions on the scarce resource is to do a cost-benefit analysis. This type of analysis contrasts different options of resource allocation and, based on an evaluation of the expected courses of action and the consequences of these actions, the optimal way to do so in the light of different policy goals can be elicited.

Benefit-cost analysis basically can look at several ways of solving a problem and then assigning the best route for a solution, based on the set of consequences that would result from the further development of the individual courses of action, and then choosing the course of action that results in the least amount of damage to the expected outcome for the environmental quality that remains after that development or process takes place. Further complicating this analysis are the interrelationships of the various parts of the environment that might be impacted by the chosen course of action. Sometimes it is almost impossible to predict the various outcomes of a course of action, due to the unexpected consequences and the amount of unknowns that are not accounted for in the benefit-cost analysis.

Energy

Sustainable energy is clean and can be used over a long period of time. Unlike fossil fuels and biofuels that provide the bulk of the worlds energy, renewable energy sources like hydroelectric, solar and wind energy produce far less pollution. Solar energy is commonly used on public parking meters, street lights and the roof of buildings. Wind power has expanded quickly, its share of worldwide electricity usage at the end of 2014 was 3.1%. Most of California's fossil fuel infrastructures are sited in or near low-income communities, and have traditionally suffered the most from California's fossil fuel energy system. These communities are historically left out during the decision-making process, and often end up with dirty power plants and other dirty energy projects that poison the air and harm the area. These toxicants are major contributors to health problems in the communities. As renewable energy becomes more common, fossil fuel infrastructures are replaced by renewables, providing better social equity to these communities. Overall, and in the long run, sustainable development in the field of energy is also deemed to contribute to economic sustainability and national security of communities, thus being increasingly encouraged through investment policies.

Manufacturing

Technology

One of the core concepts in sustainable development is that technology can be used to assist people meet their developmental needs. Technology to meet these sustainable development needs is often referred to as appropriate technology, which is an ideological movement (and its manifestations) originally articulated as intermediate technology by the economist E. F. Schumacher in his influential work, Small is Beautiful. and now covers a wide range of technologies. Both Schumacher and many modern-day proponents of appropriate technology also emphasise the technology as people-centered. Today appropriate technology is often developed using open source principles, which have led to open-source appropriate technology (OSAT) and thus many of the plans of the technology can be freely found on the Internet. OSAT has been proposed as a new model of enabling innovation for sustainable development.

Transport

Transportation is a large contributor to greenhouse gas emissions. It is said that one-third of all gasses produced are due to transportation. Motorized transport also releases exhaust fumes that contain particulate matter which is hazardous to human health and a contributor to climate change.

Sustainable transport has many social and economic benefits that can accelerate local sustainable development. According to a series of reports by the Low Emission Development Strategies Global Partnership (LEDS GP), sustainable transport can help create jobs, improve commuter safety through investment in bicycle lanes and pedestrian pathways, make access to employment and social opportunities more affordable and efficient. It also offers a practical opportunity to save people's time and household income as well as government budgets, making investment in sustainable transport a 'win-win' opportunity.

Some western countries are making transportation more sustainable in both long-term and short-term implementations. An example is the modifications in available transportation in Freiburg, Germany. The city has implemented extensive methods of public transportation, cycling, and walking, along with large areas where cars are not allowed.

Since many western countries are highly automobile-orientated areas, the main transit that people use is personal vehicles. About 80% of their travel involves cars. Therefore, California, is one of the highest greenhouse gases emitters in the United States. The federal government has to come up with some plans to reduce the total number of vehicle trips in order to lower greenhouse gases emission. Such as:

- Improve public transit through the provision of larger coverage area in order to provide more mobility and accessibility, new technology to provide a more reliable and responsive public transportation network.

- Encourage walking and biking through the provision of wider pedestrian pathway, bike share station in commercial downtown, locate parking lot far from the shopping center, limit on street parking, slower traffic lane in downtown area.

- Increase the cost of car ownership and gas taxes through increased parking fees and tolls, encouraging people to drive more fuel efficient vehicles. This can produce a social equity problem, since lower income people usually drive older vehicles with lower fuel efficiency. Government can use the extra revenue collected from taxes and tolls to improve public transportation and benefit poor communities.

Other states and nations have built efforts to translate knowledge in behavioral economics into evidence-based sustainable transportation policies.

Business

The most broadly accepted criterion for corporate sustainability constitutes a firm's efficient use of natural capital. This eco-efficiency is usually calculated as the economic value added by a firm in relation to its aggregated ecological impact. This idea has been popularised by the World Business Council for Sustainable Development (WBCSD) under the following definition: "Eco-efficiency is achieved by the delivery of competitively priced goods and services that satisfy human needs and bring quality of life, while progressively reducing ecological impacts and resource intensity throughout the life-cycle to a level at least in line with the earth's carrying capacity." (DeSimone and Popoff, 1997: 47)

Similar to the eco-efficiency concept but so far less explored is the second criterion for corporate sustainability. Socio-efficiency describes the relation between a firm's value added and its social impact. Whereas, it can be assumed that most corporate impacts on the environment are negative (apart from rare exceptions such as the planting of trees) this is not true for social impacts. These can be either positive (e.g. corporate giving, creation of employment) or negative (e.g. work accidents, mobbing of employees, human rights abuses). Depending on the type of impact socio-efficiency thus either tries to minimise negative social impacts (i.e. accidents per value added) or maximise positive social impacts (i.e. donations per value added) in relation to the value added.

Both eco-efficiency and socio-efficiency are concerned primarily with increasing economic sustainability. In this process they instrumentalise both natural and social capital aiming to benefit from win-win situations. However, as Dyllick and Hockerts point out the business case alone will not be sufficient to realise sustainable development. They point towards eco-effectiveness, socio-effectiveness, sufficiency, and eco-equity as four criteria that need to be met if sustainable development is to be reached.

CASI Global, New York "CSR & Sustainability together lead to sustainable development. CSR as in corporate social responsibility is not what you do with your profits, but is the way you make profits. This means CSR is a part of every department of the company value chain and not a part of HR / independent department. Sustainability as in effects towards Human resources, Environment and Ecology has to be measured within each department of the company.

Income

At the present time, sustainable development, along with the solidarity called for in Catholic social teaching, can reduce poverty. While over many thousands of years the 'stronger' (economically or physically) overcame the weaker, nowadays for various reasons - Catholic social teaching, social solidarity, sustainable development – the stronger helps the weaker. This aid may take various forms. 'The Stronger' offers real help rather than striving for the elimination or annihilation of the other. Sustainable development reduces poverty through financial (among other things, a balanced budget), environmental (living conditions), and social (including equality of income) means.

Architecture

In sustainable architecture the recent movements of New Urbanism and New Classical architecture promote a sustainable approach towards construction, that appreciates and develops smart growth, architectural tradition and classical design. This in contrast to modernist and International Style architecture, as well as opposing to solitary housing estates and suburban sprawl, with long commuting distances and large ecological footprints. Both trends started in the 1980s. (It should be noted that sustainable architecture is predominantly relevant to the economics domain while architectural landscaping pertains more to the ecological domain.)

Politics

A study concluded that social indicators and, therefore, sustainable development indicators, are scientific constructs whose principal objective is to inform public policy-making. The International Institute for Sustainable Development has similarly developed a political policy framework, linked to a sustainability index for establishing measurable entities and metrics. The framework consists of six core areas, international trade and investment, economic policy, climate change and energy, measurement and assessment, natural resource management, and the role of communication technologies in sustainable development.

The United Nations Global Compact Cities Programme has defined sustainable political development in a way that broadens the usual definition beyond states and governance. The political is defined as the domain of practices and meanings associated with basic issues of social power as they pertain to the organisation, authorisation, legitimation and regulation of a social life held in common. This definition is in accord with the view that political change is important for responding to economic, ecological and cultural challenges. It also means that the politics of economic change can be addressed. They have listed seven subdomains of the domain of politics:

1. Organization and governance

2. Law and justice

3. Communication and critique

4. Representation and negotiation

5. Security and accord

6. Dialogue and reconciliation

7. Ethics and accountability

This accords with the Brundtland Commission emphasis on development that is guided by human rights principles .

Culture

Working with a different emphasis, some researchers and institutions have pointed out that a fourth dimension should be added to the dimensions of sustainable development, since the triple-bottom-line dimensions of economic, environmental and social do not seem to be enough to reflect the complexity of contemporary society. In this context, the Agenda 21 for culture and the United Cities and Local Governments (UCLG) Executive Bureau lead the preparation of the policy statement "Culture: Fourth Pillar of Sustainable Development", passed on 17 November 2010, in the framework of the World Summit of Local and Regional Leaders – 3rd World Congress of UCLG, held in Mexico City. This document inaugurates a new perspective and points to the relation between culture and sustainable development through a dual approach: developing a solid cultural policy and advocating a cultural dimension in all public policies. The Circles of Sustainability approach distinguishes the four domains of economic, ecological, political and cultural sustainability.

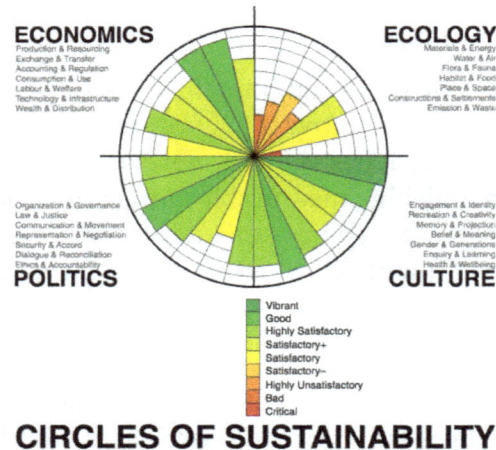

CIRCLES OF SUSTAINABILITY

Framing of sustainable development progress according to the Circles of Sustainability, used by the United Nations.

Other organizations have also supported the idea of a fourth domain of sustainable development. The Network of Excellence "Sustainable Development in a Diverse World", sponsored by the European Union, integrates multidisciplinary capacities and interprets cultural diversity as a key element of a new strategy for sustainable development. The Fourth Pillar of Sustainable Development Theory has been referenced by executive director of IMI Institute at UNESCO Vito Di Bari in his manifesto of art and architectural movement Neo-Futurism, whose name was inspired by the 1987 United Nations' report Our Common Future. The Circles of Sustainability approach used by Metropolis defines the (fourth) cultural domain as practices, discourses, and material expressions, which, over time, express continuities and discontinuities of social meaning.

Themes

Progress

The United Nations Conference on Sustainable Development (UNCSD; also known as Rio 2012) was the third international conference on sustainable development, which aimed at reconciling the economic and environmental goals of the global community. An outcome of this conference was the development of the Sustainable Development Goals that aim to promote sustainable progress and eliminate inequalities around the world. However, few nations met the World Wide Fund for Nature's definition of sustainable development criteria established in 2006. Although some nations are more developed than others, all nations are constantly developing because each nation struggles with perpetuating disparities, inequalities and unequal access to fundamental rights and freedoms.

Measurement

In 2007 a report for the U.S. Environmental Protection Agency stated: "While much discussion and effort has gone into sustainability indicators, none of the resulting systems clearly tells us whether our society is sustainable. At best, they can tell us that we are heading in the wrong direction, or that our current activities are not sustainable. More often, they simply draw our attention to the existence of problems, doing little to tell us the origin of those problems and nothing to tell us how to solve them." Nevertheless, a majority of authors assume that a set of well defined and harmonised indicators is the only way to make sustainability tangible. Those indicators are expected to be identified and adjusted through empirical observations (trial and error).

Deforestation and increased road-building in the Amazon Rainforest are a concern because of increased human encroachment upon wilderness areas, increased resource extraction and further threats to biodiversity.

The most common critiques are related to issues like data quality, comparability, objective function and the necessary resources. However a more general criticism is coming from the project management community: How can a sustainable development be achieved at global level if we cannot monitor it in any single project?

The Cuban-born researcher and entrepreneur Sonia Bueno suggests an alternative approach that is based upon the integral, long-term cost-benefit relationship as a measure and monitoring tool for the sustainability of every project, activity or enterprise. Furthermore, this concept aims to be a practical guideline towards sustainable development following the principle of conservation and increment of value rather than restricting the consumption of resources.

Reasonable qualifications of sustainability are seen U.S. Green Building Council's (USGBC) Leadership in Energy and Environmental Design (LEED). This design incorporates some ecological, economic, and social elements. The goals presented by LEED design goals are sustainable sites, water efficiency, energy and atmospheric emission reduction, material and resources efficiency, and indoor environmental quality. Although amount of structures for sustainability development is many, these qualification has become a standard for sustainable building.

Recent research efforts created also the SDEWES Index to benchmark the performance of cities across aspects that are related to energy, water and environment systems. The SDEWES Index consists of 7 dimensions, 35 indicators, and close to 20 sub-indicators. It is currently applied to 58 cities.

Natural Capital

The sustainable development debate is based on the assumption that societies need to manage three types of capital (economic, social, and natural), which may be non-substitutable and whose consumption might be irreversible. Leading ecological economist and steady-state theorist Herman Daly, for example, points to the fact that natural capital can not necessarily be substituted by economic capital. While it is possible that we can find ways to replace some natural resources, it is much more unlikely that they will ever be able to replace eco-system services, such as the protection provided by the ozone layer, or the climate stabilizing function of the Amazonian forest. In fact natural capital, social capital and economic capital are often complementarities. A further obstacle to substitutability lies also in the multi-functionality of many natural resources. Forests, for example, not only provide the raw material for paper (which can be substituted quite easily), but they also maintain biodiversity, regulate water flow, and absorb CO_2.

Deforestation of native rain forest in Rio de Janeiro City for extraction of clay for civil engineering (2009 picture).

Another problem of natural and social capital deterioration lies in their partial irreversibility. The loss of biodiversity, for example, is often definitive. The same can be true for cultural diversity. For example, with globalisation advancing quickly the number of indigenous languages is dropping at alarming rates. Moreover, the depletion of natural and social capital may have non-linear consequences. Consumption of natural and social capital may have no observable impact until a certain threshold is reached. A lake can, for example, absorb nutrients for a long time while actually increasing its productivity. However, once a certain level of algae is reached lack of oxygen causes the lake's ecosystem to break down suddenly.

Business-as-usual

Before flue-gas desulfurization was installed, the air-polluting emissions from this power plant in New Mexico contained excessive amounts of sulfur dioxide.

If the degradation of natural and social capital has such important consequence the question arises why action is not taken more systematically to alleviate it. Cohen and Winn point to four types of market failure as possible explanations: First, while the benefits of natural or social capital depletion can usually be privatised, the costs are often externalised (i.e. they are borne not by the party responsible but by society in general). Second, natural capital is often undervalued by society since we are not fully aware of the real cost of the depletion of natural capital. Information asymmetry is a third reason—often the link between cause and effect is obscured, making it difficult for actors to make informed choices. Cohen and Winn close with the realization that contrary to economic theory many firms are not perfect optimisers. They postulate that firms often do not optimise resource allocation because they are caught in a "business as usual" mentality.

Education

Education must be revisited in light of a renewed vision of sustainable human and social development that is both equitable and viable. This vision of sustainability must take into consideration the social, environmental and economic dimensions of human development and the various ways in which these relate to education: 'An empowering education is one that builds the human resources we need to be productive, to continue to learn, to solve problems, to be creative, and to live together and with nature in peace and harmony. When nations ensure that such an education is accessible to all throughout their lives, a quiet revolution is set in motion: education becomes the engine of sustainable development and the key to a better world.'

Higher education in sustainability across education streams including engineering, finance, supply chain and operations is gaining weight-age. Multiple institutes including Wharton, Columbia, CASI Global New York offer certifications in Sustainability. Corporate's prefer employees certified in sustainability.

Insubstantial Stretching of the Term

It has been argued that since the 1960s, the concept of sustainable development has changed from 'conservation management' to 'economic development', whereby the original meaning of the concept has been stretched somewhat.

In the 1960s, the international community realised that many African countries needed national plans to safeguard wildlife habitats, and that rural areas had to confront the limits imposed by soil, climate and water availability. This was a strategy of conservation management. In the 70s, however, the focus shifted to the broader issues of the provisioning of basic human needs, community participation as well as appropriate technology use throughout the developing countries (and not just in Africa). This was a strategy of economic development, and the strategy was carried even further by the The Brundtland Commission's report on Our Common Future when the issues went from regional to international in scope and application. In effect, the conservationists were crowded out and superseded by the developers.

But shifting the focus of sustainable development from conservation to development has had the imperceptible effect of stretching the original forest management term of sustainable yield from the use of renewable resources only (like forestry), to now also accounting for the use of non-renewable resources (like minerals). This stretching of the term has been questioned. Thus, environmental economist Kerry Turner has argued that literally, there can be no such thing as overall 'sustainable development' in an industrialised world economy that remains heavily dependent on the extraction of Earth's finite stock of exhaustible mineral resources:

> "It makes no sense to talk about the sustainable use of a non-renewable resource (even with substantial recycling effort and use rates). Any positive rate of exploitation will eventually lead to exhaustion of the finite stock. "

In effect, it has been argued that the Industrial Revolution as a whole is unsustainable.

One critic has argued that the Brundtland Commission promoted nothing but a business as usual strategy for world development, with the ambiguous and insubstantial concept of 'sustainable development' attached as a public relations slogan. The report on Our Common Future was largely the result of a political bargaining process involving many special interest groups, all put together to create a common appeal of political acceptability across borders. After World War II, the notion of 'development' had been established in the West to imply the projection of the American model of society onto the rest of the world. In the 1970s and 1980s, this notion was broadened somewhat to also imply human rights, basic human needs and finally, ecological issues. The emphasis of the report was on helping poor nations out of poverty and meeting the basic needs of their growing populations — as usual. This issue demanded more economic growth, also in the rich countries, who would then import more goods from the poor countries to help them out — as usual. When the discussion switched to global ecological limits to growth, the obvious dilemma was left aside by calling for economic growth with improved resource efficiency, or what was termed 'a change in the quality of growth'. However, most countries in the West had experienced such improved resource efficiency since the early 20th century already and as usual; only, this improvement had been more than offset by continuing industrial expansion, to the effect that world resource consumption was now higher than ever before — and these two historical trends were completely ignored in the report. Taken together, the policy of perpetual economic growth for the entire planet remained virtually intact. Since the publication of the report, the ambiguous and insubstantial slogan of 'sustainable development' has marched on worldwide, the critic concludes.

Need for Sustainable Development

The great surge in development and technology over the last two centuries has contributed to the increase in quantities of chemicals sufficient to damage natural systems on a grand scale, more often than not, involving the whole world. Acid rain, desertification, destruction of species, greenhouse modification and ozone layer depletion are well known examples of these impacts.

Protection of the natural ecosystem has long-term benefits for humans in utilitarian terms through maintenance of gene pools, bio-diversity and other potentially useful factors and, in spiritual terms, through living in harmony with nature. The ecosystem's intrinsic values and rights, regardless of human needs, therefore, should be taken into account apart from considering it a resource to be exploited for human settlement, food and energy production.

Sustainable development, therefore, is imperative. It is defined as a pattern of social and structured economic transformation (i.e., development), which optimises the economic and societal benefits available in the present, without jeopardising the likely potential for similar benefits in the future. A primary goal of sustainable development is to achieve a reasonable and equitably distributed level of economic well being that can be perpetuated continuously for many human generations. This implies using natural resources in a renewable manner that does not eliminate or degrade them, or diminish their usefulness for future generations. It further implies using non-renewable (i.e., exhaustible) mineral resources in a manner that does not unnecessarily preclude easy access to them by future generations.

Three of the several definitions that capture these aspects are given below:

(i) Development that meets the needs of the present without compromising the ability of future generations to meet their own needs (WCED, 1987).

(ii) Sustainable development ensures that the maximum rate of resource consumption and waste discharge for a selected development portfolio would be sustained indefinitely, in a defined planning region, without progressively impairing its bio-productivity and ecological integrity. Environmental conservation, therefore, contrary to general belief, accelerates rather than hinders economic development. Therefore, the development plans have to ensure:

1. Sustainable and equitable use of resources for meeting the needs of the present and future generations without causing damage to environment;

2. To prevent further damage to our life-support systems;

3.To conserve and nurture the biological diversity, gene pool and other resources for long-term food security (MoEF, 1999).

(iii) The primary objective of the Sustainable Development is to reduce the absolute poverty of the world's poor through providing lasting and secure livelihoods that minimize resource depletion, environmental degradation, cultural disruption and social instability

Thus, the doctrine of sustainability provides a mechanism for developments to occur in harmony with environmental protection and enhancement. It presents a challenge to technologists to manage construction and manufacturing procedures so as to cause no net environmental harm and to preserve environmental resources for future generations.

Some of the approaches to EM to achieve sustainability include the following:

Ad hoc approach: This means an approach developed in reaction to a specific situation.

Problem-solving approach: This refers to an approach that follows a series of logical steps to identify problems, needs and solutions.

Systems approach: This approach focuses on ecosystem (e.g., mountain; high latitude; savannah; desert; island; lake, etc.), agro-ecosystem, etc.

Regional approach: This is based on ecological zones or biogeophysical units such as watershed, river basin, coastal zone, islands, etc.

Specialist discipline approach: This refers to the approach often adopted by professionals like air quality management, water quality management, land management, environmental health, urban management, conservation area management, etc.

Voluntary sector approach: This represents the approach NGOs (Non-Government Organisations) encourage and/or support.

Em Tools

One of the outcomes of the three decades of policy development since 1970 has been the evolution of the techniques for the analysis and management of environmental effects. The tools considered in this Course include:

(i) Environment assessment (EA).

(ii) Economic assessment, usually through cost-benefit analysis (CBA).

(iii) Environmental Impact Statement (EIS).

(iv) Environment audits.

(v) Waste minimisation programs and environmental management systems.

(vi) Life cycle assessments (LCA).

(vii) Environmental design (ED).

These tools are indispensable as far as environmental management is concerned. But, rather than being comprehensive, self-contained procedures, these tools are really aids that can be adapted for various situations. Legislative and bureaucratic frameworks for environmental management differ considerably, not only between nations, but also among states or regions within countries. Some procedures fit within others, for example, an environment impact statement is a tool or instrument, used within a more general environment assessment process. The methods are contributions to decision-making, rather than complete decision-making procedures.

Participants In Em

Adam (1990) identified the following two groups involved in environment and development:

(i) People or governments who are uninformed of the implications of development, or who are unable to voice their views adequately and affect change.

(ii) Consultants, scientists, economists and bankers and those bent on riches or blinkered by concern for sovereignty, religion or national security.

In any given environmental management situation there are likely to be a number of different perspectives, and hence the various possible responses. An environmental manager has to grasp the sum total of the perspectives and try to avoid conflicts between participants and minimise damage to the environment. The participants of environmental management can be categorised as existing users, groups seeking change, groups with little control, the public, facilitators and controllers.

Existing Users

This category refers to those who currently use land or other resources. And, those using the environment or resources usually evolve rights and develop management skills. However, problems arise where unwritten traditional strategies and rights break down or get usurped, typically, by incoming migrants and settlers, urban elites or powerful commercial organisations. Worldwide, the expropriation of common resources from traditional users has become a problem (The Ecologist, 1993).

There has, however, been a growing practice of seeking to consult and involve local people (i.e., indigenous groups) in environmental management, and to understand and make wider use of indigenous knowledge (Klee, 1980). EM can learn a lot from the study of local people's livelihood strategies.

Since the 1975 - 1985 UN Decade for Women, there has been an interest in studying the role of women in EM. According to the different perspectives adopted, these studies take different shapes as under:

- Women, environment and development: This focuses on women as having a special relationship with the environment as its users and managers.

- Gender and development: This seeks gender as a key dimension of social difference affecting people's experiences, concerns and capabilities.

- Women in development: This focuses on reasons for women's exclusion or marginalisation from decision-making and receipt of the benefits of development.

Groups Seeking Change

This category consists of governments (with conflicting demands from various ministries or policy-makers), commerce (e.g., national, local, multi-national companies, etc.), individuals seeking personal gain or seeking to change the prevailing situation, international agencies, NGOs, media, academics, etc. It is probably the exception to the rule for special-interest groups not to control policymaking and development, although a few do so with the aim of improving environmental care. The environmental manager should be vigilant for such control, and seek to reduce it if it acts against environmental quality. When environmental management involves more than one country, which is often the case, negotiation skills are at least as important as access to technology, knowledge and management strategies (Vogler and Imber, 1995).

Groups with Little Control

This category consists of the poor with no option but to overexploit what is available without investing in improvement; refugees, migrants, relocates, eco-refugees (i.e., those forced to move or marginalised so that they change the environment to survive), workers in industry/mining, etc., who face health and safety challenges, while carrying out changes. Many identify poverty alleviation and environmental care as two challenges for those in charge of development. These two issues are closely related, although linkages are often unclear and complex.

The poor, it is often claimed, degrade their environment in an effort to survive – a trap of poverty. Getting people out of poverty may be important for protecting the environment, but environmental managers must consider each local case to be sure of causes. For example, the causes of environmental degradation in urban areas may lie with policies affecting agriculturalists hundreds of kilometres away, causing them to migrate and increase city population. There are also situations where there is likely to be poverty-related environment stress: cities where population growth is outstripping employment and infrastructure; marginal, often vulnerable land where people have relocated, areas where traditional livelihood strategies are degenerating.

The Public

People, who are affected as bystanders, may wish to develop, conserve or change practices (if aware of what is happening) and those out of global concern form part of this category. The public usually consists of more than one group of people who probably have different, perhaps conflicting, views and goals with powerful groups dominating the situation. Environmental managers must, therefore, establish the needs of the weak and ensure that they are not ignored, yet work with the influential.

Sustainable development strategies need to be designed to fit local conditions and to be co-ordinated to ensure that one locality does not conflict with another. Environmental management should act as mediator and catalyst to develop collaborative approaches (Selin and Chavez, 1995). Advantages of public involvement in environmental management are as follows:

(i) The public may be able to provide advice on management considering local conditions.

(ii) Often planning and management should be more accountable and more careful.

(iii) Fears and opposition to management may be reduced, if people are informed.

(iv) The communication gap between the experts and locals can be reduced.

Facilitators

This category consists of funding agencies, consultants, planners, workers including migrants affected by health and safety issues, etc.

Funding bodies can support environmentally desirable developments or withhold money until proposals are modified to meet required standards. Starting with the World Bank in the early 1970s, most funding bodies have developed environmental management units, guidelines and manuals (Turnham, 1991). There is also a huge diversity of bodies conducting research aimed at improving environmental management: universities, private research companies, independent international research institutes and UN or UN-related agencies. Most research is applied in response to perceived needs, but some is anticipatory and warns about possible threats and potentially useful strategies.

Controllers

This category consists of government and international agencies, traditional rulers and religions, planners, law, consumer protection bodies and NGOs (including various green/environmental bodies), trade organisations, media, concerned individuals, academics, global opinion, environmental managers, etc.

The skills of environmental managers and ecologists are vital to determine the best strategies for the survival of fauna and flora and to organise sustainable land and resource use. NGOs have become important watchdogs of corporate, government and special-interest group activities. They have a multifaceted role: lobbying at international meetings and at national government level; media campaigning to increase public awareness and empowerment; fund-raising for environmental management, conservation and environmental education; researching environmentally sound strategies and approaches; acting as ginger groups to identify environmental problems and fight for their control, etc.

Ethics And The Environment

''Environmental ethics is the discipline in philosophy that studies the moral relationship of human beings, and also the value and moral status of, the environment and its nonhuman contents''..

It is concerned with the issue of responsible personal conduct with respect to natural landscapes, resources, species, and non-human organisms. It is cluster of beliefs, values and norms regarding how humans should interact with the environment.

Human effects on the environment today have consequences for the future, and therefore, discussions of environmental ethics also involve the rights of future generations. The arguments for and against various principles in environmental ethics are made more complex because of conflicting values. The resolution of the resulting conflict requires that we recognize differing values and have a basic scientific knowledge about the environment as well as the ability to clearly formulate a logical argument.

The study would make you understand topics such as, the environmental philosophy with illustration of Aldo Leopold's land ethic and factors that demand environmental ethics.

The Environmental philosophy

Environmental philosophy is that wing of philosophy that expresses trepidation with natural environment and livelihood of humans. Main areas of interest for philosophers include defining environment and its value, environmentalism and deep ecology, endangered species and restoration of nature. Its major components are environmental ethics, theology, environmental aesthetics and ecofeminism.

Aldo Leopold, (Annexure I) formulated ecological restoration focusing on Land ethic in a book „A Sand County Almanac, 1949' (Annexure II), defined a new link between nature and people and has a stage for modern conservation movement. „For embracing this ethic ecologically literate citizens are required who can also solve global environmental challenges.

"This Land ethic simply enlarges the boundaries of the community to include soils, waters, plants, and animals, or collectively: the land."

Environmental ethics and factors that necessitate it:

In early 1970's philosopher's formulated environmental ethics as a study concerned with the value of the physical and biological environment. The focus of this study contrasts with traditional ethical studies, which had to do with the relationships among people. There are utilitarian, ecological, aesthetic and moral reasons for placing a value on the environment.

The major factors that necessitate environmental ethics are:

(i) New effects on nature: Because our modern technological civilisation affects nature greatly, we must examine the ethical consequences of these new actions.

(ii) New knowledge about nature: Modern science demonstrates how we have changed and are changing our environment in ways not previously understood, thus raising new ethical issues. For example, until the past decade, few people believed that human activities could be changing the Earth"s global environment. Now, however, scientists believe that burning fossil fuels and clearing forests have changed the amount of carbon dioxide in the atmosphere, and this causes changes in our climates, and hence the need for global action. This new perspective raises new moral issues.

(iii) Expanding moral concerns: Some people argue that animals, trees and even rocks have moral and legal rights, and that it is a natural extension of civilisation to begin including the environment in ethics. These expanded concerns lead to need for new ethics.

After knowing Leopold's philosophy and ethics of environment there is a need for us to understand the history behind a diverse scientific, social and political movement for addressing environmental issues.

Environmental Movement

Apollo 8's Earthrise, 24 December 1968

The environmental movement (sometimes referred to as the ecology movement), also including conservation and green politics, is a diverse scientific, social, and political movement for addressing environmental issues. Environmentalists advocate the sustainable management of resources and stewardship of the environment through changes in public policy and individual behavior. In its recognition of humanity as a participant in (not enemy of) ecosystems, the movement is centered on ecology, health, and human rights.

The environmental movement is an international movement, represented by a range of organizations, from the large to grassroots and varies from country to country. Due to its large membership, varying and strong beliefs, and occasionally speculative nature, the environmental movement is not always united in its goals. The movement also encompasses some other movements with a more specific focus, such as the climate movement. At its broadest, the movement includes private citizens, professionals, religious devotees, politicians, scientists, nonprofit organizations and individual advocates.

History

Early Awareness

Levels of air pollution rose during the Industrial Revolution, sparking the first modern environmental laws to be passed in the mid-19th century

Early interest in the environment was a feature of the Romantic movement in the early 19th century. The poet William Wordsworth had travelled extensively in the Lake District and wrote that it is a "sort of national property in which every man has a right and interest who has an eye to perceive and a heart to enjoy".

The origins of the environmental movement lay in the response to increasing levels of smoke pollution in the atmosphere during the Industrial Revolution. The emergence of great factories and the concomitant immense growth in coal consumption gave rise to an unprecedented level of air pollution in industrial centers; after 1900 the large volume of industrial chemical discharges added to the growing load of untreated human waste. Under increasing political pressure from the urban middle-class, the first large-scale, modern environmental laws came in the form of Britain's Alkali Acts, passed in 1863, to regulate the deleterious air pollution (gaseous hydrochloric acid) given off by the Leblanc process, used to produce soda ash.

Conservation Movement

The modern conservation movement was first manifested in the forests of India, with the practical application of scientific conservation principles. The conservation ethic that began to evolve included three core principles: that the human activity damaged the environment, that there was a civic duty to maintain the environment for future generations, and that scientific, empirically based methods should be applied to ensure this duty was carried out. Sir James Ranald Martin was prominent in promoting this ideology, publishing many medico-topographical reports that demonstrated the scale of damage wrought through large-scale deforestation and desiccation, and lobbying extensively for the institutionalization of forest conservation activities in British India through the establishment of Forest Departments. The Madras Board of Revenue started local conservation efforts in 1842, headed by Alexander Gibson, a professional botanist who systematically adopted a forest conservation program based on scientific principles. This was the first case of state management of forests in the world. Eventually, the government under Governor-General Lord Dalhousie introduced the first permanent and large-scale forest conservation program in the world in 1855, a model that soon spread to other colonies, as well the United States. In 1860, the Department banned the use shifting cultivation. Dr. Hugh Cleghorn's 1861 manual, The forests and gardens of South India, became the definitive work on the subject and was widely used by forest assistants in the subcontinent.

Students from the forestry school at Oxford, on a visit to the forests of Saxony in the year 1892

Sir Dietrich Brandis joined the British service in 1856 as superintendent of the teak forests of Pegu division in eastern Burma. During that time Burma's teak forests were controlled by militant Karen tribals. He introduced the "taungya" system, in which Karen villagers provided labour for clearing, planting and weeding teak plantations. He formulated new forest legislation and helped establish research and training institutions. The Imperial Forest School at Dehradun was founded by him.

Formation of Environmental Protection Societies

The late 19th century saw the formation of the first wildlife conservation societies. The zoologist Alfred Newton published a series of investigations into the Desirability of establishing a 'Close-time' for the preservation of indigenous animals between 1872 and 1903. His advocacy for legislation to protect animals from hunting during the mating season led to the formation of the Plumage League (later the Royal Society for the Protection of Birds) in 1889. The society acted as a protest group campaigning against the use of great crested grebe and kittiwake skins and feathers in fur clothing. The Society attracted growing support from the suburban middle-classes, and influenced the passage of the Sea Birds Preservation Act in 1869 as the first nature protection law in the world.

For most of the century from 1850 to 1950, however, the primary environmental cause was the mitigation of air pollution. The Coal Smoke Abatement Society was formed in 1898 making it one of the oldest environmental NGOs. It was founded by artist Sir William Blake Richmond, frustrated with the pall cast by coal smoke. Although there were earlier pieces of legislation, the Public Health Act 1875 required all furnaces and fireplaces to consume their own smoke.

John Ruskin an influential thinker who articulated the Romantic ideal of environmental protection and conservation

Systematic and general efforts on behalf of the environment only began in the late 19th century; it grew out of the amenity movement in Britain in the 1870s, which was a reaction to industrialization, the growth of cities, and worsening air and water pollution. Starting with the formation of the Commons Preservation Society in 1865, the movement championed rural preservation against the encroachments of industrialisation. Robert Hunter, solicitor for the society, worked with Hardwicke Rawnsley, Octavia Hill, and John Ruskin to lead a successful campaign to prevent the construction of railways to carry slate from the quarries, which would have ruined the unspoilt valleys of Newlands and Ennerdale. This success led to the formation of the Lake District Defence Society (later to become The Friends of the Lake District).

In 1893 Hill, Hunter and Rawnsley agreed to set up a national body to coordinate environmental conservation efforts across the country; the "National Trust for Places of Historic Interest or Natural Beauty" was formally inaugurated in 1894. The organisation obtained secure footing through the 1907 National Trust Bill, which gave the trust the status of a statutory corporation. and the bill was passed in August 1907.

An early "Back-to-Nature" movement, which anticipated the romantic ideal of modern environmentalism, was advocated by intellectuals such as John Ruskin, William Morris, and Edward Carpenter, who were all against consumerism, pollution and other activities that were harmful to the natural world. The movement was a reaction to the urban conditions of the industrial towns, where sanitation was awful, pollution levels intolerable and housing terribly cramped. Idealists championed the rural life as a mythical Utopia and advocated a return to it. John Ruskin argued that people should return to a small piece of English ground, beautiful, peaceful, and fruitful. We will have no steam engines upon it . . . we will have plenty of flowers and vegetables . . . we will have some music and poetry; the children will learn to dance to it and sing it.

Practical ventures in the establishment of small cooperative farms were even attempted and old rural traditions, without the "taint of manufacture or the canker of artificiality", were enthusiastically revived, including the Morris dance and the maypole.

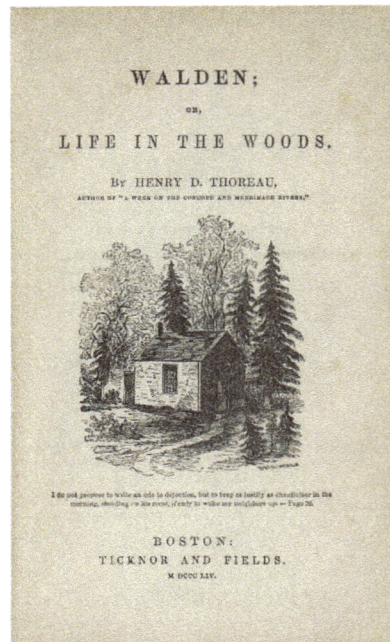

Original title page of Walden by Henry David Thoreau

The movement in the United States began in the late 19th century, out of concerns for protecting the natural resources of the West, with individuals such as John Muir and Henry David Thoreau making key philosophical contributions. Thoreau was interested in peoples' relationship with nature and studied this by living close to nature in a simple life. He published his experiences in the book Walden, which argues that people should become intimately close with nature. Muir came to believe in nature's inherent right, especially after spending time hiking in Yosemite Valley and studying both the ecology and geology. He successfully lobbied congress to form Yosemite National Park and went on to set up the Sierra Club in 1892. The conservationist principles as well as the belief in an inherent right of nature were to become the bedrock of modern environmentalism. However, the early movement in the U.S. developed with a contradiction; preservationists like John Muir wanted land and nature set aside for its own sake, and conservationists, such as Gifford Pinchot (appointed as the first Chief of the US Forest Service from 1905-1910), wanted to manage natural resources for human use.

20th Century

In the 20th century, environmental ideas continued to grow in popularity and recognition. Efforts were starting to be made to save some wildlife, particularly the American bison. The death of the last passenger pigeon as well as the endangerment of the American bison helped to focus the minds of conservationists and popularize their concerns. In 1916 the National Park Service was founded by US President Woodrow Wilson. Pioneers of the movement called for more efficient and professional management of natural resources. They fought for reform because they believed the destruction of forests, fertile soil, minerals, wildlife and water resources would lead to the downfall of society. The group that has been the most active in recent years is the climate movement.

The conservation of natural resources is the fundamental problem. Unless we solve that problem, it will avail us little to solve all others.

The U.S movement did not really take off until after World War II as people began to recognize the costs of environmental negligence, disease, and widespread air and water pollution through the occurrence of several environmental disasters that occurred post-World War II. Aldo Leopold wrote "A Sand County Almanac" in the 1940s. He believed in a land ethic that recognized that maintaining the "beauty, integrity, and health of natural systems" as a moral and ethical imperative.

Another important book in the promotion of the environmental movement was Rachel Carson's "Silent Spring" about declining bird populations due to DDT, an insecticide, pollution and man's attempts to control nature through use of synthetic substances. Her core message for her readers, was to identify the complex and fragile ecosystem and the threats facing the people. In 1958 Carson started to work on her last book, with an idea that nature needs human protection. Her influence was radioactive fallout, smog, food additives, and pesticide use. Carson's main focus was on pesticides, which led her to identify nature ad fragile and the use of technology dangerous to humans and other species.

Both of these books helped bring the issues into the public eye Rachel Carson's "Silent Spring" sold over two million copies.

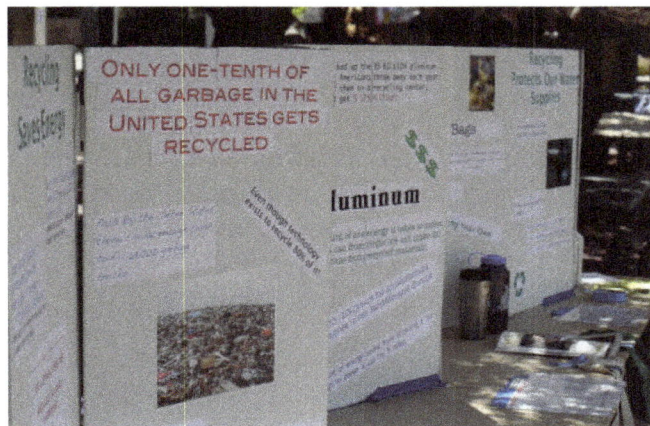

Earth Day 2007 at City College, San Diego

The first Earth Day was celebrated on 22 April 1970. Its founder, former Wisconsin Senator, Gaylord Nelson was inspired to create this day of environmental education and awareness after seeing the oil spill off the coast of Santa Barbara in 1969. Greenpeace was created in 1971 as an organization that believed that political advocacy and legislation were ineffective or inefficient solutions and supported non-violent action. 1980 saw the creation of Earth First!, a group with an ecocentric view of the world – believing in equality between the rights of humans to flourish, the rights of all other species to flourish and the rights of life-sustaining systems to flourish.

In the 1950s, 1960s, and 1970s, several events illustrated the magnitude of environmental damage caused by humans. In 1954, a hydrogen bomb test at Bikini Atoll exposed the 23 man crew of the Japanese fishing vessel Lucky Dragon 5 to radioactive fallout. In 1967 the oil tanker Torrey Canyon ran aground off the coast of Cornwall, and in 1969 oil spilled from an offshore well in California's Santa Barbara Channel. In 1971, the conclusion of a lawsuit in Japan drew international attention to the effects of decades of mercury poisoning on the people of Minamata.

At the same time, emerging scientific research drew new attention to existing and hypothetical threats to the environment and humanity. Among them were Paul R. Ehrlich, whose book The Population Bomb (1968) revived Malthusian concerns about the impact of exponential population growth. Biologist Barry Commoner generated a debate about growth, affluence and "flawed technology." Additionally, an association of scientists and political leaders known as the Club of Rome published their report The Limits to Growth in 1972, and drew attention to the growing pressure on natural resources from human activities.

Meanwhile, technological accomplishments such as nuclear proliferation and photos of the Earth from outer space provided both new insights and new reasons for concern over Earth's seemingly small and unique place in the universe.

In 1972, the United Nations Conference on the Human Environment was held in Stockholm, and for the first time united the representatives of multiple governments in discussion relating to the state of the global environment. This conference led directly to the creation of government environmental agencies and the UN Environment Program.

By the mid-1970s anti-nuclear activism had moved beyond local protests and politics to gain a wider appeal and influence. Although it lacked a single co-ordinating organization the anti-nuclear movement's efforts gained a great deal of attention, especially in the United Kingdom and United States. In the aftermath of the Three Mile Island accident in 1979, many mass demonstrations took place. The largest one was held in New York City in September 1979 and involved 200,000 people.

Since the 1970s, public awareness, environmental sciences, ecology, and technology have advanced to include modern focus points like ozone depletion, global climate change, acid rain, and the potential of genetically modified organisms (GMOs). Some argue that genetically modified plants and animals are inherently bad because they are unnatural. Others point out the possible benefits of GM crops such as water conservation through corn modified to be less "thirsty" and decreased pesticide use through insect - resistant crops.

United States

Beginning in the conservation movement at the beginning of the 20th century, the contemporary environmental movement's roots can be traced back to Murray Bookchin's Our Synthetic Environment, Paul R. Ehrlich's The Population Bomb, and Rachel Carson's Silent Spring. American environmentalists have campaigned against nuclear weapons and nuclear power in 1960s and 1970s, acid rain in the 1980s, ozone depletion and deforestation in the 1990s, and most recently climate change and global warming.

The United States passed many pieces of environmental legislation in the 1970s, such as the Clean Water Act, the Clean Air Act, the Endangered Species Act, and the National Environmental Policy Act. These remain as the foundations for current environmental standards.

Timeline of Us Environmental History

- 1832- Hot Springs Reservation
- 1864- Yosemite Valley

- 1872- Yellowstone National Park

- 1892- Sierra Club

- 1916- National Park Service Organic Act

- 1916- National Audubon Society

- 1949- UN Scientific Conference on the Conservation and Utilization of Resources

- 1961- World Wildlife Foundation

- 1964- Land and Water Conservation Act

- 1964- National Wilderness Preservation System

- 1968- National Trails System Act

- 1968- National Wild and Scenic Rivers System/Wild and Scenic Rivers Act

- 1969- National Environmental Policy Act

- 1970- First Earth Day- 22 April

- 1970- Clean Air Act

- 1970- Environmental Protection Agency

- 1971- Greenpeace

- 1972- Clean Water Act

- 1973- Endangered Species Act

- 1980- Earth First!

- 1992- UN Earth Summit in Rio de Janeiro

Latin America

After the International Environmental Conference in Stockholm in 1972 Latin American officials returned with a high hope of growth and protection of the fairly untouched natural resources. Governments spent millions of dollars, and created departments and pollution standards. However, the outcomes have not always been what officials had initially hoped. Activists blame this on growing urban populations and industrial growth. Many Latin American countries have had a large inflow of immigrants that are living in substandard housing. Enforcement of the pollution standards is lax and penalties are minimal; in Venezuela, the largest penalty for violating an environmental law is 50,000 bolivar fine ($3,400) and 3 days in jail. In the 1970s or 1980s many Latin American countries were transitioning from military dictatorships to democratic governments.

Brazil

In 1992, Brazil came under scrutiny with the United Nations Conference on Environment and Development in Rio de Janeiro. Brazil has a history of little environmental awareness. It has the highest biodiversity in the world and also the highest amount of habitat destruction. One-third of the world's forests lie in Brazil, and they have the largest river, The Amazon, and the largest rainforest, the Amazon Rainforest. The people have raised funds to create state parks and increase the consciousness of people who have destroyed forests and polluted waterways. They have several organizations that have fronted the environmental movement. The Blue Wave Foundation was created in 1989 and has partnered with advertising companies to promote national education campaigns to keep Brazil's beaches clean. Funatura was created in 1986 and is a wildlife sanctuary program. Pro-Natura International is a private environmental organization created in 1986.

Europe

In 1952 the Great London Smog episode killed thousands of people and led the UK to create the first Clean Air Act in 1956. In 1957 the first major nuclear accident occurred in Windscale in northern England. The supertanker Torrey Canyon ran aground off the coast of Cornwall in 1967 causing the first major oil leak that killed marine life along the coast. In 1972, in Stockholm, the United Nations Conference on the Human Environment created the UN Environment Programme. The EU's environmental policy was formally founded by a European Council declaration and the first five-year environment programme was adopted. The main idea of the declaration was that prevention is better than the cure and the polluter should pay. 1979 saw the partial meltdown of Three Mile Island in the USA.

In the 1980s the green parties that were created a decade before began to have some political success.. In 1986, there was a nuclear accident in Chernobyl, Ukraine. The end of the 1980s and start of the 1990s saw the fall of communism across central and Eastern Europe, the fall of the [Berlin Wall], and the Union of East and West Germany. In 1992 there was a UN summit held in Rio de Janeiro where Agenda 21 was adopted. The Kyoto Protocol was created in 1997 which set specific targets and deadlines to reduce global greenhouse gas emissions. In the early 2000s activists believed that environmental policy concerns were overshadowed by energy security, globalism, and terrorism.

Asia

Middle East

The environmental movement is reaching the less developed world with different degrees of success. The Arab world, including the Middle East and North Africa, has different adaptations of the environmental movement. Countries on the Persian Gulf have high incomes and rely heavily on the large amount of energy resources in the area. Each country in the Arab world has varying combinations of low or high amounts of natural resources and low or high amounts of labor.

The League of Arab States has one specialized sub-committee, of 12 standing specialized subcommittee in the Foreign Affairs Ministerial Committees, which deals with Environmental Issues. Countries in the League of Arab States have demonstrated an interest in environmental issue, on paper some environmental activists have doubts about the level of commitment to environmental issues;; being a part of the world community may have obliged these countries to portray concern for the environment. Initial level of environmental awareness may be the creation of a ministry of the environment. The year of establishment of a ministry is also indicative of level of engagement. Saudi Arabia was the first to establish environmental law in 1992 followed by Egypt in 1994. Somalia is the only country without environmental law. In 2010 the Environmental Performance Index listed Algeria as the top Arab country at 42 of 163; Morocco was at 52 and Syria at 56. The Environmental Performance Index measures the ability of a country to actively manage and protect their environment and the health of their citizens. A weighted index is created by giving 50% weight for environmental health objective (health) and 50% for ecosystem vitality (ecosystem); values range from 0-100. No Arab countries were in the top quartile, and 7 countries were in the lowest quartile.

South Korea and Taiwan

South Korea and Taiwan experienced similar growth in industrialization from 1965-1990 with few environmental controls. South Korea's Han River and Nakdong River were so polluted by unchecked dumping of industrial waste that they were close to being classified as biologically dead. Taiwan's formula for balanced growth was to prevent industrial concentration and encourage manufacturers to set up in the countryside. This led to 20% of the farmland being polluted by industrial waste and 30% of the rice grown on the island was contaminated with heavy metals. Both countries had spontaneous environmental movements drawing participants from different classes. Their demands were linked with issues of employment, occupational health, and agricultural crisis. They were also quite militant; the people learned that protesting can bring results. The polluting factories were forced to make immediate improvements of the conditions or pay compensation to victims. Some were even forced to shut down or move locations. The people were able to force the government to come out with new restrictive rules on toxins, industrial waste, and air pollution. All of these new regulations caused the migration of those polluting industries from Taiwan and South Korea to China and other countries in Southeast Asia with more relaxed environmental laws.

China

China's environmental movement is characterized by spontaneous alliances that often only occur at the local level. The Chinese have realized the ability of riots and protests to have success and had led to an increase in disputes in China by 30% since 2005 to more than 50,000 events. Protests cover topics such as environmental issues, land loss, income, and political issues. They have also grown in size from about 10 people or fewer in the mid-1990s to 52 people per incident in 2004. China has more relaxed environmental laws than other countries in Asia, so many polluting factories have relocated to China causing pollution in China. Water pollution, water scarcity, soil pollution, soil degradation, and desertification are issues currently in discussion in China. The groundwater table of the North China Plain is dropping by 1.5 m (5 ft) per year. This groundwater table occurs in the region of China that produces 40% of the country's grain.

India

Environmental and public health is an ongoing struggle within India. The first seed of an environmental movement in India was the foundation in 1964 of Dasholi Gram Swarajya Sangh, a labour coperative started by Chandi Prasad Bhatt. It was inaugurated by Sucheta Kriplani and founded on a land donated by Shyma Devi. This initiative was eventually followed up with the Chipko movement starting in 1974.

The most severe single event underpinning the movement was the Bhopal gas leakage on 3 December 1984. 40 tons of methyl isocyanate was released, immediately killing 2,259 people and ultimately affecting 700,000 citizens.

India has a national campaign against Coca-Cola and Pepsi Cola plants due to their practices of drawing ground water and contaminating fields with sludge. The movement is characterized by local struggles against intensive aquaculture farms. The most influential part of the environmental movement in India is the anti-dam movement. Dam creation has been thought of as a way for India to catch up with the West by connecting to the power grid with giant dams, coal or oil-powered plants, or nuclear plants. Jhola Aandolan a mass movement is conducting as fighting against polyethylene carry bags uses and promoting cloth/jute/paper carry bags to protect environment & nature. Activists in the Indian environmental movement consider global warming, sea levels rising, and glaciers retreating decreasing the amount of water flowing into streams to be the biggest challenges for them to face in the early twenty-first century. Eco Revolution movement has been started by Eco Needs Foundation in 2008 from Aurangabad Maharashtra state .The pioneer of Eco Revolution movement is Priyanand Agale more than 20,000 youths associated with this movement to strengthen this movement Eco Needs Foundation organized mass environmental awareness programmes. To sought participation of children,Youths, researchers, spiritual and political leaders.Foundation had conducted International conferences at India Eco Revolution 2011 which concluded with Aurangabad Declaration for River Conservation. Eco Revolution 2012 conference was conducted in Sri Lanka which concluded with Colombo Declaration on spirituality for environmental conservation in collaboration with Sri Lanka government. Eco Revolution 2013 was conducted at Nepal which concluded with the Phokhara Declaration for effect of climate change at high altitude .Foundation launched the worlds first environmental social networking site ecoface.in. Foundation developed model of sustainable development at Dhanora village of Dholpur, Rajasthan as a India's First smart village.

Bangladesh

Mithun Roy Chowdhury, President, Save Nature & Wildlife (SNW), Bangladesh, insisted that the people of Bangladesh raise their voice against Tipaimukh Dam, being constructed by the Government of India. He said Tipaimukh Dam project will be another "death trap for Bangladesh like the Farakka Barrage," that would lead to an environmental disaster for 50 million people in the Meghna River basin. He said that this project will start desertification in Bangladesh.

Scope of the Movement

Environmental science is the study of the interactions among the physical, chemical and biological components of the environment.

Before flue-gas desulfurization was installed, the air-polluting emissions from this power plant in New Mexico contained excessive amounts of sulfur dioxide

Ecology, or ecological science, is the scientific study of the distribution and abundance of living organisms and how these properties are affected by interactions between the organisms and their environment.

Primary Focus Points

The environmental movement is broad in scope and can include any topic related to the environment, conservation, and biology, as well as preservation of landscapes, flora, and fauna for a variety of purposes and uses. See List of environmental issues. When an act of violence is committed against someone or some institution in the name of environmental defense it is referred to as eco terrorism.

- The conservation movement seeks to protect natural areas for sustainable consumption, as well as traditional (hunting, fishing, trapping) and spiritual use.

- Environmental conservation is the process in which one is involved in conserving the natural aspects of the environment. Whether through reforestation, recycling, or pollution control, environmental conservation sustains the natural quality of life.

- Environmental health movement dates at least to Progressive Era, and focuses on urban standards like clean water, efficient sewage handling, and stable population growth. Environmental health could also deal with nutrition, preventive medicine, aging, and other concerns specific to human well-being. Environmental health is also seen as an indicator for the state of the environment, or an early warning system for what may happen to humans

- Environmental justice is a movement that began in the U.S. in the 1980s and seeks an end to environmental racism and prevent low-income and minority communities from an unbalanced exposure to highways, garbage dumps, and factories. The Environmental Justice movement seeks to link "social" and "ecological" environmental concerns, while at the same time preventing de facto racism, and classism. This makes it particularly adequate for the construction of labor-environmental alliances.

- Ecology movement could involve the Gaia Theory, as well as Value of Earth and other interactions between humans, science, and responsibility.

- Bright green environmentalism is a currently popular sub-movement, which emphasizes the idea that through technology, good design and more thoughtful use of energy and resources, people can live responsible, sustainable lives while enjoying prosperity.

- Light green, and dark green environmentalism are yet other sub-movements, respectively distinguished by seeing environmentalism as a lifestyle choice (light greens), and promoting reduction in human numbers and/or a relinquishment of technology (dark greens)

- Deep Ecology is an ideological spinoff of the ecology movement that views the diversity and integrity of the planetary ecosystem, in and for itself, as its primary value.

- The anti-nuclear movement opposes the use of various nuclear technologies. The initial anti-nuclear objective was nuclear disarmament and later the focus began to shift to other issues, mainly opposition to the use of nuclear power. There have been many large anti-nuclear demonstrations and protests. Major anti-nuclear groups include Campaign for Nuclear Disarmament, Friends of the Earth, Greenpeace, International Physicians for the Prevention of Nuclear War, and the Nuclear Information and Resource Service.

Environmental Law and Theory

Property Rights

Many environmental lawsuits question the legal rights of property owners, and whether the general public has a right to intervene with detrimental practices occurring on someone else's land. Environmental law organizations exist all across the world, such as the Environmental Law and Policy Center in the midwestern United States.

Citizens' Rights

One of the earliest lawsuits to establish that citizens may sue for environmental and aesthetic harms was Scenic Hudson Preservation Conference v. Federal Power Commission, decided in 1965 by the Second Circuit Court of Appeals. The case helped halt the construction of a power plant on Storm King Mountain in New York State.

Nature's Rights

Christopher D. Stone's 1972 essay, "Should trees have standing?" addressed the question of whether natural objects themselves should have legal rights. In the essay, Stone suggests that his argument is valid because many current rightsholders (women, children) were once seen as objects.

Environmental Reactivism

Numerous criticisms and ethical ambiguities have led to growing concerns about technology, including the use of potentially harmful pesticides, water additives like fluoride, and the extremely dangerous ethanol-processing plants.

NIMBY syndrome refers to public outcry caused by knee-jerk reaction to an unwillingness to be exposed to even necessary developments. Some serious biologists and ecologists created the scientific ecology movement which would not confuse empirical data with visions of a desirable future world.

Environmentalism Today

Today, the sciences of ecology and environmental science, in addition to any aesthetic goals, provide the basis of unity to some of the serious environmentalists. As more information is gathered in scientific fields, more scientific issues like biodiversity, as opposed to mere aesthetics, are a concern to environmentalists. Conservation biology is a rapidly developing field.

In recent years, the environmental movement has increasingly focused on global warming as one of the top issues. As concerns about climate change moved more into the mainstream, from the connections drawn between global warming and Hurricane Katrina to Al Gore's film An Inconvenient Truth, more and more environmental groups refocused their efforts. In the United States, 2007 witnessed the largest grassroots environmental demonstration in years, Step It Up 2007, with rallies in over 1,400 communities and all 50 states for real global warming solutions.

Composite images of Earth generated by NASA in 2001 (left) and 2002 (right)

Many religious organizations and individual churches now have programs and activities dedicated to environmental issues. The religious movement is often supported by interpretation of scriptures. Most major religious groups are represented including Jewish, Islamic, Anglican, Orthodox, Evangelical, Christian and Catholic.

Radical environmentalism

Radical environmentalism emerged from an ecocentrism-based frustration with the co-option of mainstream environmentalism. The radical environmental movement aspires to what scolar Christopher Manes calls "a new kind of environmental activism: iconoclastic, uncompromising, discontented with traditional conservation policy, at times illegal ..." Radical environmentalism presupposes a need to reconsider Western ideas of religion and philosophy (including capitalism, patriarchy and globalization) sometimes through "resacralising" and reconnecting with nature. Greenpeace represents an organisation with a radical approach, but has contributed in serious ways towards understanding of critical issues, and has a science-oriented core with radicalism as a means to media exposure. Groups like Earth First! take a much more radical posture. Some radical environmentalist groups, like Earth First! and the Earth Liberation Front, illegally sabotage or destroy infrastructural capital.

Criticisms

Conservative critics of the movement characterize it as radical and misguided. Especially critics of the United States Endangered Species Act, which has come under scrutiny lately, and the Clean Air Act, which they said conflict with private property rights, corporate profits and the nation's overall economic growth. Critics also challenge the scientific evidence for global warming. They argue that the environmental movement has diverted attention from more pressing issues.

International Environmental Movement

Environmental impacts of developmental activities are being sounded since the early 1960s. One of the first and most influential warnings was Silent Spring, a book about the use of pesticides, written by Rachel Carson and published in 1962. In the mid-1960s some scientists were sounding warnings about the possibility of climate change due to increased carbon dioxide levels in the atmosphere resulting from burning of fossil fuels. The same period witnessed early environmental conferences like the two British "Countryside in 1970" conferences of 1963 and 1965, and the U.S. White House conference on "Natural Beauty" of 1965.

In the United States of America (USA), this climate of thought and innovative legislators produced a law that represented a landmark in environmental management not only in the United States but worldwide – the National Environmental Policy Act (NEPA), 1969. Revise The NEPA required "in agency planning and decision making an integrated interdisciplinary use of the natural and social services and the environmental design arts". It also required agencies to take account of non-quantified values, and to use ecological information and concepts in planning and decisionmaking affecting the quality of the environment.

"Spaceship Earth" arrived in the 1970s with concern growing, some warnings, and a piece of new style landmark legislation to be added to the many pieces of air and water pollution control law that existed already around the globe. The 1970s was an era of response and action. The most influential responses were the United Nations Conferences on the Environment (1972), Population (1974) and Human Settlements (1976).

The UN Conference was held in June 1972 in Stockholm, Sweden. As a result, the United Nations Environment programme (UNEP) was set up together with a fund to finance major projects. A 26-point declaration of environmental principles was adopted, calling for commitments by countries to deal with environmental problems of international significance. Approval was given to the Earth-Watch Program – a co-ordinated plan to use and expand existing monitoring systems to measure pollution levels around the world and their effects on climate. A convention to control ocean dumping of shore-generated wastes was endorsed. The International Whaling Commission was urged to adopt a ten-year moratorium on commercial whaling. Other proposals for conservation conventions were endorsed (e.g., the World Heritage Trust to protect areas of unique natural, historical or cultural value that are part of the heritage of all humankind, and a convention to protect endangered species of plants and animals, etc.).

Three months prior to the U N Conference a report on phase one of the Club of Rome's projects on the Predicament of Mankind was published. The Limits to Growth described the work on a global model of the five factors that determine and limit growth – population, agricultural production natural resources, industrial production and pollution. The Club of Rome first came together in 1968 out of the concern of thirty individuals from ten countries to discuss the present and future predicament of humanity and to foster understanding of the interdependent components that make up the global system.

Another conference, less visible than that of the United Nations, but like Limits to Growth, something as pointer to future, occurred also in 1972. This was the Second World Conference on National Parks and Protected Areas, held under the aegis of the World Conservation Union (IUCN). IUCN was established in 1948, and its work and influence would become of increasing importance, as biodiversity and the still accelerating forces of population growth, technology and pollution reduced the gene stocks of wild nature.

(In 1972 world population was 3.7 billion.) At this time the now well-known conflicts and concerns about the loss of tropical forests were yet to come. At this time also, conservation and development were seen as opposing, rather than complementary objectives as inferred by UNCED.

In parallel with, and following the Stockholm Conference, there was a surge of policy-making and institution building in the developed world. In the decade of the 1970s, many governments introduced environmental legislation, established agencies with environmental responsibilities, or grafted these onto existing departments. Others moved more cautiously, with "pilot" policies that could be elaborated after experience had been gained. Environmental impact assessment (EIA), the discipline that could steer decision-making towards allowance for environmental factors was introduced in some countries. Often regarded with suspicion as an unnecessary impediment to legitimate developmental objectives and progress, EIA took a decade to be acknowledged as a tool that could actually produce projects superior in both quality and value.

In 1971, the World Bank established an environmental section to promote environmental reconnaissance of hydro-projects in recognition of one of the fundamentals of environmental management. The principle is that; knowledge of the environmental context of development proposals should be a base within which planning proceeds with constant reference to potential impacts. The major negative impacts of some hydroprojects had already included spread of water-borne disease. Loss of biodiversity due to destruction of habitat and wildlife, water quality effects on reservoirs and downstream locations, induced seismicity and decline of productive ecosystems. One of the objectives of the World Bank initiative was capacity building. Bank agents may have carried out the initial reconnaissance but borrowers were aided and encouraged towards developing competence and setting up of in-house environmental units.

Development economic growth, technological expansion and exponential population increase continued and problems of global scale surfaced. By 1980, the handling and disposal of hazardous wastes was recognised as a national and global issue. Through the 1970s and 1980s, the growth of chemical manufacturing in developing countries exceeded that of the developed world. Samples showed that developing world populations ranked high in exposure to toxic chemicals, blood lead levels and DDT contamination of human milk. Effective national regulation and control programmes existed in developed countries, but not elsewhere. The discovery of grossly contaminated sites (a danger to water supplies, food chains, and human health generally) in the developed world could only raise questions, but not answers, about the developing world where disposal is haphazard, and neither disposal facilities nor national control frameworks exist.

In I970, most air pollution legislation treated air pollution as a local phenomenon. By 1980, it had become global and was affecting the forests and lakes of Europe and North America. The major cause was the burning of fossil fuels. Global emissions of sulphur dioxides grew by an estimated 470% and emissions of carbon dioxide tenfold, in the 20th century. Nitrous oxide emissions in the USA have increased nine times over the same period. Given suitable atmospheric conditions, sulphur dioxides and nitrous oxides can be transported long distances and transformed into acids. It was realised that acid rain affected not only lakes and streams but also crops and vegetation.

At ground level, ozone is also damaging to crops and vegetation. But in the upper atmosphere ozone acts as a filter by absorbing harmful wavelengths of ultra-violet radiation. In 1985, it was confirmed that the ozone shield over the Antarctic was thinning, and a "hole" had developed, and an unprecedented, international action followed for a world action plan. Its main components were global monitoring, estimating the impact of changes in the ozone layer on radiation, skin cancer, ecosystems and regional climate, and collecting data on production and emissions. While the world action plan was developing, an international convention was being hammered out. The Montreal Protocol, providing a framework for action by each country, was agreed in 1987. Officials from most of the CFC producing/using countries agreed to a 50% reduction by 1999. But, new scientific evidence indicated that the situation was more serious than had been thought. The Helsinki Declaration of 1989 stated the intention of 80 countries to completely phase out CFCs by the year 2000.

Another, and greater, global concern was also growing as the ozone saga unfolded. The concerns about increasing CO_2 levels, expressed by a few scientists in 1960s were confirmed in the 1980s by a large proportion of the scientific community. Data showed increasing concentrations, not only of CO_2 but also nitrous oxide (NO_2), methane (CH_4) and specific chlorofluorocarbons. Their addition to the atmosphere permits it to absorb more of the infrared radiation emitted from the earth.

The understanding and application of the concept that environment and development are integrated, complementary and indivisible, and are to be thought of as one, is fundamental to environmental management. This new message was given in 1980 in the World Conservation Strategy, published by the World Conservation Union (IUCN), the United Nations Environment Programme (UNEP) and the World Wide Fund for nature (WWF). Its successor document, Caring for the Earth, was published in 1991. This strategy, so clearly complement to Our Common

Future sets out the rationale and the strategic actions needed to achieve a sustainable future.

The original World Conservation Strategy of 1980 introduced an important innovation in environmental management – the national level conservation strategy, subsequently prepared by over 50 countries. These identify the country"s most urgent environmental needs, assist decision-makers in determining priorities, allocate resources and build institutional capacity to handle complex environmental issues. Preparation includes fundamental reexamination of laws, policies and institution. Similar to these national conservation strategies were national environmental action plans sponsored by the World Bank.

In the early 1980s, the then Minister of the Environment of the Government of Norway, Dr. Gro Harlem Brundtland worried by the growing incompatibility between human development on Earth and the possibility for Nature to withstand the resulting ecological stress undertook to launch a global investigation of this incompatibility. With the assistance of a former Vice President of the Republic of Sudan, she managed to assemble a team of 26 concerned persons of high calibre into what was named the World Commission on Environment and Development (WCED). The WCED project was given the green light - but not a full endorsement - by the United Nations and succeeded in obtaining support in kind and cash from a good number of countries. There were some noticeable exceptions to the supporting countries, such as the USA, Japan, France, the United Kingdom and Germany.

Nevertheless, the WCED had sufficient means to commission a wide range of necessary surveys and studies by competent and well known scientists and professionals, access the statistics of many countries and all international organizations, and undertake a series of hearings in a sample of countries around the world. Its exploration of the issues ended sometime in 1986. With the help of some countries (e.g., Canada), the WCED undertook to prepare a summary of its findings, with a synthesis, conclusions and recommendations. The result was a paperback report, under the title Our Common Future, published in April 1987 by Oxford University Press, gradually translated into a large number of languages and distributed widely around the world. The report was officially presented to the general Assembly of the United Nations in November 1987, and enthusiastically approved by most, if not all members.

At the core of the recommendations made by the Commission was the goal of sustainable development, which was defined as development that meets the needs of the present without compromising the ability of future generations to meet their own needs.

Our Common Future was a precursor, though not the only one, to United Nation"s Conference on Environment and Development (UNCED) of June 1992 (popularly known as the Rio Earth summit). Both the report and the Conference link environment and development, bringing together within the concept of systems and interdependence, environment and development objectives once considered being in opposition. And, in 1992, Agenda 21, a global plan of action for sustainable development was adopted. In addition to adopting Agenda 21, the assembled leaders from over 100 countries signed the Framework Convention on Climate Change and the Convention on Biological Diversity and endorsed the Rio Declaration and the Forest Principles.

The Rio Declaration represents a set of 27 agreed principles aimed at the objective of a new and suitable global partnership, international agreements that respect the integrity of the global environmental and developmental system, and recognition of the integral and interdependent nature of Earth, our home.

The outcomes of the discussion of Agenda 21 occupy more than 450 pages. Agenda 21 is the operational plan for moving humankind into the age of sustainability. It should be read and given profound consideration by all people, especially all technologists. Its implementation will demand commitment and the capacity of all nations, people and individuals.

More than half the sections of Agenda 21 are of direct or indirect relevance to science and technology, e.g., those dealing with the protection of human health, human settlement, integration of environment and development in decision-making, protection of the atmosphere, integrated approaches to the planning and management of land resources, combating desertification and approaches to planning and management of land resources, combating desertification and drought, sustainable mountain development, conservation of biological diversity, environmentally sound management of biotechnology, toxic chemicals, hazardous wastes, solid wastes, radioactive wastes, the scientific and technological community, transfer of environmentally sound technology, education, national mechanisms, institutional arrangements and information.

The Commission on Sustainable Development (CSD) was created in December 1992 to ensure an effective follow-up of UNCED; to monitor and report on implementation of the Earth Summit agreements at the local, national, regional and international levels. The CSD is a functional commission of the UN Economic and Social Council (ECOSOC), with 53 members. It was agreed that a five-year review of Earth Summit progress would be made in 1997 by the United Nations General Assembly meeting in special session.

Assembly held in June 1997 adopted a comprehensive document entitled Programme for the Further Implementation of Agenda 21 prepared by CSD. It also adopted the programme of work of the Commission for 1998 - 2002.

The World Summit on Sustainable Development took place in Johannesburg, South Africa, 2 - 11 September, 2002. Thousands participated at the Johannesburg Summit: heads of State and Government, national delegates and representatives from business and industry, children and youth, farmers, indigenous people, local authorities, non-governmental organisations, scientific and technological communities, women and workers and trade unions.

According to General Assembly resolution 55/199 of 20 December 2000, the review would focus on the identification of accomplishments and areas where further efforts are needed to implement Agenda 21 and other outcomes of the UNCED. At the national level, Governments are encouraged to prepare national assessments on the progress achieved and challenges encountered. At the regional level, the UN Department of Economic and Social Affairs (UNDESA) have sponsored round tables of eminent persons and UN regional commissions facilitated regional ministerial preparatory meetings. At the global level, the UN Secretariat for the Summit organised three substantive meetings in the first half of the year 2002 before convening the Summit in September.

Against the backdrop of the discussions we have had so far, we will next focus on the major environmental concerns in India.

Environmental Concerns In India

We can categorise environmental concerns in India as those arising from negative effects of the very process of development; and conditions of poverty and under development. The major problems, which encompass the area of EM, are:

- Burgeoning population and its impact on life support systems, which negate the positive impacts of the developmental programmes.

- Rapid growth in the population of domesticated animals over the past few decades accompanied by a loss of area under grassland and pastures. Around 3.5% of the geographical area is under grasslands, while the domesticated animal population numbers nearly 500 million (1991).

- Out of a total area of about 329 million hectares, 175 million hectares of land require special treatment to restore them for productive and profitable use. Land degradation is caused by water and wind erosion (159 million ha), salinity and alkalinity (8 million ha), and river action and other factors (7 million ha).

- The forest wealth is dwindling due to overgrazing and over-exploitation both for commercial and household needs, encroachments, unsustainable practices including certain practices of shifting cultivation, and developmental activities such as roads, buildings, irrigation and power projects.

- The recorded forest cover in the country is about 75 million hectare, which is 19.5% of the total geographical area against the national goal of 33% in the plains and 66% for hilly regions. Even within this area, a meagre 11% constitutes forests with 40% or more of crown cover. The annual rate of loss of the forest cover is 47,500 hectares.

- The loss of habitat is leading to the extinction of plant, animal and microbial species. Over 1500 plant and animal species are in the endangered category.

- The wetlands of India, which are rich in aquatic and bird life providing food and shelter as also the breeding and spawning ground for the marine and fresh water fishes, are facing problems of pollution and over-exploitation.

- The major rivers of the country are facing problems of pollution and siltation. The coastline is under stress and coastal areas have been severely damaged due to indiscriminate construction, waste disposal near the water-line and aquaculture. Coastal vegetation including mangroves and sea grasses is facing extinction. The mountain ecosystems are under threat of serious degradation.

- India is witnessing a rising requirement for forest based goods due which there is extensive deforestation leading to severe loss of natural resources and in turn the erosion of valuable topsoil, is threatening the livelihood and security of millions of hill people and also encroachment into forest protected areas. (As a result of requirement of 70 million m3 of round wood per year in India by the end of the decade and its transportation, there is fear that this could result in loss of high conservation value forests and biodiversity elsewhere)

- Increasing demand for water for different sectors such as agriculture, domestic, energy generation, industry that resulted in depletion of water source. The quality of groundwater is being affected due to chemical pollution and due to the ingress of seawater in coastal areas.

- Absence of an integrated land and water use policy for the country has resulted in a heavy toll of basic natural assets. Coral reef ecosystems are adversely affected by indiscriminate exploitation of corals for production of lime, recreational use and for ornamental trade. Island ecosystems are subjected to pressures of various forms including migration of people from the main land.

- Pollution arising from toxic wastes and non biodegradable consumer articles is on the rise. A large number of industries and other development projects are sited close to heavily populated urban centers, leading to over congestion and over pollution, as also the diversion of population and natural resources from the rural areas.

- Mechanism to indigenously arrive at a reliable estimate of total greenhouse gas emissions in the country, among various sources such as agriculture, animals, energy production and consumption, forestry and land use change, waste management, etc., is inadequate. More such reliable data, which is indigenously arrived at, is essential for negotiating international law, treaties, protocols and conventions on environment-related problems where developing countries are unduly pressurised.

It is often difficult to clearly delineate the causes and consequences of environmental degradation in terms of simple cause effect relationships. The causes and effects are often interwoven in complex webs of socio-economic, technological and implementation factors.

The liberalisation of economy in India in the recent past is expected to promote consumerism by throwing a wide array of products with wide options to the consumers. This, in turn, is expected to put pressure on the natural resources in the form of raw materials, and therefore, may deplete the natural resources and cause irreversible damages. Several plant and animal species may become extinct and the non-renewable resources may be exhausted soon depriving the future generations of their benefits. This has been amply witnessed in the past in several western countries where industrial revolution depleted the natural resources for generations to come. Today, these countries are looking towards the developing countries for raw material and natural resources. India should learn a lesson from the past experiences of industrial development within India and outside of it.

References

- Johnson, D.L., S.H. Ambrose, T.J. Bassett, M.L. Bowen, D.E. Crummey, J.S. Isaacson, D.N. Johnson, P. Lamb, M. Saul, and A.E. Winter-Nelson. 1997. Meanings of environmental terms. Journal of Environmental Quality 26: 581–589

- Huesemann, Michael H., and Joyce A. Huesemann (2011). Technofix: Why Technology Won't Save Us or the Environment, Chapter 6, "Sustainability or Collapse?", New Society Publishers, ISBN 0865717044

- The California Institute of Public Affairs (CIPA) (August 2001). "An ecosystem approach to natural resource conservation in California". CIPA Publication No. 106. InterEnvironment Institute. Retrieved 10 July 2012

- Karamanos, P., Voluntary Environmental Agreements: Evolution and Definition of a New Environmental Policy Approach. Journal of Environmental Planning and Management, 2001. 44(1): p. 67-67-84

- Australian achievements in environment protection and nature conservation 1972-1982. Canberra: Australian Environment Council and Council of Nature Conservation Ministers. 1982. pp. 1–2. ISBN 0-642-88655-5

- Paul Sattler and Colin Creighton. "Australian Terrestrial Biodiversity Assessment 2002". National Land and Water Resources Audit. Department of Sustainabililty, Environment, Water, Population and Communities. Retrieved 21 September 2011

- Kneese, Allen V.; Ayres, Robert U.; D'Arge, Ralph C. (1970). Economics and the environment: a materials balance approach. Resources for the Future; distributed by the Johns Hopkins Press, Baltimore. ISBN 978-0-8018-1215-6

- Jessica Andersson; Daniel Slunge (16 June 2005). "Tanzania – Environmental Policy Brief" (PDF). Tanzania – Environmental Policy Brief. Development Partners Group Tanzania. Retrieved 10 July 2012

- Guo Z, Zhang L, Li Y (2010). "Increased dependence of humans on ecosystem services and biodiversity". PLoS ONE. 5 (10). doi:10.1371/journal.pone.0013113. PMC 2948508 . PMID 20957042

- Daly, Herman E.; Cobb, John B. Jr (1994). For The Common Good: Redirecting the Economy toward Community, the Environment, and a Sustainable Future. Beacon Press. ISBN 978-0-8070-4705-7

- IPCC AR4 SYR (2007). Core Writing Team; Pachauri, R.K; Reisinger, A., eds. Climate Change 2007: Synthesis Report. Contribution of Working Groups I, II and III to the Fourth Assessment Report of the Intergovernmental Panel on Climate Change. IPCC. pp. 1–22. ISBN 92-9169-122-4

- Circles of Sustainability Urban Profile Process and Scerri, Andy; James, Paul (2010). "Accounting for sustainability: Combining qualitative and quantitative research in developing 'indicators' of sustainability". International Journal of Social Research Methodology. 13 (1): 41–53. doi:10.1080/13645570902864145

- United Nations Environment Programme (1978). Holling, C.S., ed. Adaptive environmental assessment and management. International Institute for Applied Systems Analysis. ISBN 978-0-471-99632-3

- Ministério do Meio Ambiente (2012). "Ministério do Meio Ambiente". Ministério do Meio Ambiente (in Portuguese). Ministério do Meio Ambiente. Retrieved 10 July 2012

- Armitage, D.R.; Berkes, F.; Doubleday, N. (2007). Adaptive Co-Management: Collaboration, Learning, and Multi-Level Governance. Vancouver: UBC Press. ISBN 978-0-7748-1383-9

- "Collaborative Australian Protected Areas Database 2008". Department of Sustainability, Environment, Water, Population and Communities. Retrieved 21 September 2011

- Zhang, S.X.; V. Babovic (2012). "A real options approach to the design and architecture of water supply systems using innovative water technologies under uncertainty" (PDF). Journal of Hydroinformatics

Environmental Management: Impact and Analysis

Environmental Impact Assessment holds a great impetus in environmental management. It refers to evaluating all the positive and negative aspects of a project or a work on the environment prior to its inauguration. The study is beneficial for project sponsors as it reduces the cost and time for implementation of the project, avoids violations of laws and treatment cost. A healthier environment, decreased resource use, etc are some of the benefits of this procedure. Environmental Impact Assessment is best understood in confluence with the major topics listed in the following section.

Environmental Systems Analysis

Environmental systems analysis (ESA) is a systematic and systems based approach for describing human actions impacting on the natural environment to support decisions and actions aimed at perceived current or future environmental problems. Impacts of different types of objects are studied that ranges from projects, programs and policies, to organizations, and products. Environmental systems analysis encompasses a family of environmental assessment tools and methods, including life cycle assessment (LCA), material flow analysis (MFA) and substance flow analysis (SFA), and environmental impact assessment (EIA), among others.

Overview

ESA studies aims at describing the environmental repercussions of defined human activities. These activities are mostly effective through use of different technologies altering material and energy flows, or (in)directly changing ecosystems (e.g. through changed land-use, agricultural practices, logging etc.), leading to undesired environmental impacts in a, more or less, specifically defined geographical area, and time, ranging from local to global. The basis for the analytical procedures used in ESA studies is the perception of flows of matter and energy associated to causal chains linking human activities to the environmental changes of concern. Some methods are focusing different parts or aspects of the energy/matter flows or the causal chains, where flow models like MFA or LCA deals with the more or less human controlled societal flows while, e.g. ecological risk assessment (ERA) is related to disentangling environmental causal chains. Environmental systems analysis studies has been suggested to be divided between "full" and "attributional" approaches. The full mode covers identified material and energy flows and associated processes leading to environmental impacts. The attributional approach, on the other hand, is based on an analysis of the processes needed to fulfil a certain purpose such as the function that a product delivers. The combination of methods (e.g. LCA and environmental risk assessment) has also been of interest

Methods can be grouped into procedural and analytical approaches. The procedural ones (e.g. EIA or strategic environmental assessment, SEA) focus on the procedure around the analysis, while the analytical ones (e.g. LCA, MFA) put the main focus on technical aspects of the analysis, and can be used as parts of the procedural approaches. Regarding the impacts studied, the environmental issues cover both effecs of natural resource use and other environmental impacts, e.g. due to emissions of chemicals, or other agents. In addition, environmental systems analysis studies can cover or be based on economic accounts (life cycle costing, cost-benefit analysis, input-output analysis, systems for economic and environmental accounts), or consider social aspects. The objects of study are distinguished into five categories. These are projects, policies and plans, regions or nations, firms and other organizations, products and functions, and substances.

Further, environmental systems analysis studies are often used to support decision making and it is acknowledged that the decision context varies and is of importance. This regards, for example, that business activities can be related in different ways to the products and other objects studied in environmental systems analysis.

History

A perception of a coherent family of tools and methods for ESA started to become established by findings published in the year 2000. Common characteristics were found to recently have appeared across tools and methods that had previously been seen as distinct from each other. The characteristics were full and attribution approaches, respectively, and the tools and methods were each earlier determined by one unique combination of flow object, spatial boundary and relation to time.

An overview of tools and methods for ESA was published five years later. It was to a large extent based on a series of reports and also drew on the life cycle management project CHAINET. The series of reports covered for example an introduction to tools for Esa that also related them to decision situations, and a study on differences and similarities between tools for esa where a short case study on heat production was included. In the CHAINET project, commissioned by the EU Environment and Climate programme, analytical tools for decision making were studied regarding demand and supply of environmental information, while procedural ESA approaches were not covered.

An expansion of the field has occurred and a number of scientific journals publish extensively on the application of ESA methods e.g. Energy and Environmental Science, Environmental Science and Technology, Journal of Cleaner Production, International Journal och Life-cycle Assessment, and Journal of Industrial Ecology.

Tools and Methods

The environmental systems analysis tools and methods include:

- Cost-benefit analysis (CBA)

- Ecological footprint (EF)

- Energy analysis (En)

- Environmental impact assessment (EIA)

- Environmental management system (EMS)

- Input-output analysis (IOA)

- Life-cycle assessment (LCA)

- Life-cycle cost analysis (LCCA)

- Material flow accounting (MFA)

- Risk assessment

- Strategic environmental assessment (SEA)

- Systems for economic and environmental accounts (SEEA)

Environmental Impact Assessment

Environmental assessment (EA) is the assessment of the environmental consequences (positive and negative) of a plan, policy, program, or concrete projects prior to the decision to move forward with the proposed action. In this context, the term "environmental impact assessment"(EIA) is usually used when applied to concrete projects by individuals or companies and the term "strategic environmental assessment" (SEA) applies to policies, plans and programmes most often proposed by organs of state. Environmental assessments may be governed by rules of administrative procedure regarding public participation and documentation of decision making, and may be subject to judicial review.

The purpose of the assessment is to ensure that decision makers consider the environmental impacts when deciding whether or not to proceed with a project. The International Association for Impact Assessment (IAIA) defines an environmental impact assessment as "the process of identifying, predicting, evaluating and mitigating the biophysical, social, and other relevant effects of development proposals prior to major decisions being taken and commitments made." EIAs are unique in that they do not require adherence to a predetermined environmental outcome, but rather they require decision makers to account for environmental values in their decisions and to justify those decisions in light of detailed environmental studies and public comments on the potential environmental impacts.

History of EIA

Environmental impact assessments commenced in the 1960s, as part of increasing environmental awareness. EIAs involved a technical evaluation intended to contribute to more objective decision making. In the United States, environmental impact assessments obtained formal status in 1969, with enactment of the National Environmental Policy Act. EIAs have been used increasingly around the world. The number of "Environmental Assessments" filed every year "has vastly overtaken the number of more rigorous Environmental Impact Statements (EIS)." An Environmental Assessment is a "mini-EIS designed to provide sufficient information to allow the agency to decide whether the preparation of a full-blown Environmental Impact Statement (EIS) is necessary." EIA is an activity that is done to find out the impact that would be done before development will occur.

Methods

General and industry specific assessment methods are available including:

- *Industrial products* - Product environmental life cycle analysis (LCA) is used for identifying and measuring the impact of industrial products on the environment. These EIAs consider activities related to extraction of raw materials, ancillary materials, equipment; production, use, disposal and ancillary equipment.

- *Genetically modified plants* - Specific methods available to perform EIAs of genetically modified organisms include GMP-RAM and INOVA.

- *Fuzzy logic* - EIA methods need measurement data to estimate values of impact indicators. However, many of the environment impacts cannot be quantified, e.g. landscape quality, lifestyle quality and social acceptance. Instead information from similar EIAs, expert judgment and community sentiment are employed. Approximate reasoning methods known as fuzzy logic can be used. A fuzzy arithmetic approach has also been proposed and implemented using a software tool (TDEIA).

Follow-up

At the end of the project, an audit evaluates the accuracy of the EIA by comparing actual to predicted impacts. The objective is to make future EIAs more valid and effective. Two primary considerations are:

- *Scientific* - to examine the accuracy of predictions and explain errors

- *Management* - to assess the success of mitigation in reducing impacts

Audits can be performed either as a rigorous assessment of the null hypothesis or with a simpler approach comparing what actually occurred against the predictions in the EIA document.

After an EIA, the precautionary and polluter pays principles may be applied to decide whether to reject, modify or require strict liability or insurance coverage to a project, based on predicted harms.

The Hydropower Sustainability Assessment Protocol is a sector specific method for checking the quality of Environmental and Social assessments and management plans.

Around the World

Australia

The history of EIA in Australia could be linked to the enactment of the U.S. National Environment Policy Act (NEPA) in 1970, which made the preparation of environmental impact statements a requirement. In Australia, one might say that the EIA procedures were introduced at a State Level prior to that of the Commonwealth (Federal), with a majority of the states having divergent views to the Commonwealth. One of the pioneering states was New South Wales, whose State Pollution Control Commission issued EIA guidelines in 1974. At a Commonwealth (Federal) level, this was followed by passing of the Environment Protection (Impact of Proposals) Act in 1974. The Environment Protection and Biodiversity Conservation Act 1999 (EPBC) superseded the Environment Protection (Impact of Proposals) Act 1974 and is the current central piece for EIA in Australia on a Commonwealth (Federal) level. An important point to note is that this Commonwealth Act does not affect the validity of the States and Territories environmental and development assessments and approvals; rather the EPBC runs as a parallel to the State/Territory Systems. Overlap between federal and state requirements is addressed via bilateral agreements or one off accreditation of state processes, as provided for in the EPBC Act.

The Commonwealth Level

The EPBC Act provides a legal framework to protect and manage nationally and internationally important flora, fauna, ecological communities and heritage places-defined in the EPBC Act as matters of 'national environmental significance'. Following are the eight matters of 'national environmental significance' to which the EPBC ACT applies:

- World Heritage sites

- National Heritage places

- RAMSAR wetlands of international significance

- Listed threatened species and ecological communities

- Migratory species protected under international agreements

- The Commonwealth marine environment

- Nuclear actions (including uranium mining)

- National Heritage.

In addition to this, the EPBC Act aims at providing a streamlined national assessment and approval process for activities. These activities could be by the Commonwealth, or its agents, anywhere in the world or activities on Commonwealth land; and activities that are listed as having a 'significant impact' on matters of 'national environment significance'.

The EPBC Act comes into play when a person (a 'proponent') wants an action (often called a 'proposal' or 'project') assessed for environmental impacts under the EPBC Act, he or she must refer the project to the Department of Environment, Water, Heritage and the Arts (Australia). This 'referral' is then released to the public, as well as relevant state, territory and Commonwealth ministers, for comment on whether the project is likely to have a significant impact on matters of national environmental significance. The Department of Environment, Water, Heritage and the Arts assess the process and makes recommendation to the minister or the delegate for the feasibility. The final discretion on the decision remains of the minister, which is not solely based on matters of 'national environmental significance' but also the consideration of social and economic impact of the project.

The Australian Government environment minister cannot intervene in a proposal if it has no significant impact on one of the eight matters of 'national environmental significance' despite the fact that there may be other undesirable environmental impacts. This is primarily due to the division of powers between the States and the Federal government and due to which the Australian Government environment minister cannot overturn a state decision.

There are strict civil and criminal penalties for the breach of EPBC Act. Depending on the kind of breach, civil penalty (maximum) may go up to $550,000 for an individual and $5.5 million for a body corporate, or for criminal penalty (maximum) of seven years imprisonment and/or penalty of $46,200.

The State and Territory Level

Australian Capital Territory (ACT)

EIA provisions within Ministerial Authorities in the ACT are found in the Chapters 7 and 8 of the *Planning and Development Act 2007* (ACT). EIA in ACT was previously administered with the help of Part 4 of the Land (Planning and Environment) Act 1991 (Land Act) and Territory Plan (plan for land-use). Note that some EIA may occur in the ACT on Commonwealth land under the EPBC Act (Cth). Further provisions of the *Australian Capital Territory (Planning and Land Management) Act 1988* (Cth) may also be applicable particularly to national land and "designated areas".

New South Wales (NSW)

In New South Wales, the Environment Planning Assessment Act 1979 (EPA) establishes three pathways for EIA. The first is under Part 5.1 of the EPAA, which provides for EIA of 'State Significant Infrastructure' projects. (From June 2011, this Part replaced Part 3A, which previously covered EIA of major projects). The second is under Part 4 of the Act dealing with development control. If a project does not require approval under Part 3A or Part 4 it is then potentially captured by the third pathway, Part 5 dealing with environment impact assessment.

Northern Territory (NT)

The EIA process in Northern Territory is chiefly administered under the Environmental Assessment Act (EAA). Although EAA is the primary tool for EIA in Northern Territory, there are further provisions for proposals in the Inquiries Act 1985 (NT).

Queensland (QLD)

There are four main EIA processes in Queensland. Firstly, under the Integrated Planning Act 1997 (IPA) for development projects other than mining. Secondly, under the Environmental Protection Act 1994 (EP Act) for some mining and petroleum activities. Thirdly, under the State Development and Public Works Organization Act 1971 (State Development Act) for 'significant projects'. Finally, Environment Protection and Biodiversity Conservation Act 1999 (Cth) for 'controlled actions'.

South Australia (SA)

The local governing tool for EIA in South Australia is the Development Act 1993. There are three levels of assessment possible under the Act in the form of an environment impact statement (EIS), a public environmental report (PER) or a Development Report (DR).

Tasmania (TAS)

In Tasmania, an integrated system of legislation is used to govern development and approval process, this system is a mixture of the Environmental Management and Pollution Control Act 1994

(EMPCA), Land Use Planning and Approvals Act 1993 (LUPAA), State Policies and Projects Act 1993 (SPPA), and Resource Management and Planning Appeals Tribunal Act 1993.

Victoria (VIC)

The EIA process in Victoria is intertwined with the Environment Effects Act 1978 and the Ministerial Guidelines for Assessment of Environmental Effects (made under the s. 10 of the EE Act).

Western Australia (WA)

The Environmental Protection Act 1986 (Part 4) provides the legislative framework for the EIA process in Western Australia. The EPA Act oversees the planning and development proposals and assesses their likely impacts on the environment.

Canada

In *Friends of the Oldman River Society v. Canada (Minister of Transportation)*, (SCC 1992) La Forest J of the Supreme Court of Canada described environmental impact assessment in terms of the proper scope of federal jurisdiction with respect to environments matters,

"Environmental impact assessment is, in its simplest form, a planning tool that is now generally regarded as an integral component of sound decision-making."

Supreme Court Justice La Forest cited (Cotton, Emond & 1981 245), "The basic concepts behind environmental assessment are simply stated: (1) early identification and evaluation of all potential environmental consequences of a proposed undertaking; (2) decision making that both guarantees the adequacy of this process and reconciles, to the greatest extent possible, the proponent's development desires with environmental protection and preservation."

La Forest referred to (Jeffrey 1989, 1.2,1.4) and (Emond 1978, p. 5) who described "...environmental assessments as a planning tool with both an information-gathering and a decision-making component" that provide "...an objective basis for granting or denying approval for a proposed development."

Justice La Forest addressed his concerns about the implications of Bill C-45 regarding public navigation rights on lakes and rivers that would contradict previous cases.(La Forest & 1973 178-80)

The Canadian Environmental Assessment Act 2012 (CEAA 2012) "and its regulations establish the legislative basis for the federal practice of environmental assessment in most regions of Canada." CEAA 2012 came into force July 6, 2012 and replaces the former *Canadian Environmental Assessment Act* (1995). EA is defined as a planning tool to identify, understand, assess and mitigate, where possible, the environmental effects of a project.

"The purposes of this Act are: (a) to protect the components of the environment that are within the legislative authority of Parliament from significant adverse environmental effects caused by a designated project; (b) to ensure that designated projects that require the exercise of a power or performance of a duty or function by a federal authority under any Act of Parliament other than this Act to be carried out, are considered in a careful and precautionary manner to avoid significant adverse environmental effects; (c) to promote cooperation and coordinated action between federal and provincial governments with respect to environmental assessments; (d) to promote communication and cooperation with aboriginal peoples with respect to environmental assessments; (e) to ensure that opportunities are provided for meaningful public participation during an environmental assessment; (f) to ensure that an environmental assessment is completed in a timely manner; (g) to ensure that projects, as defined in section 66, that are to be carried out on federal lands, or those that are outside Canada and that are to be carried out or financially supported by a federal authority, are considered in a careful and precautionary manner to avoid significant adverse environmental effects; (h) to encourage federal authorities to take actions that promote sustainable development in order to achieve or maintain a healthy environment and a healthy economy; and (i) to encourage the study of the cumulative effects of physical activities in a region and the consideration of those study results in environmental assessments."

Opposition

Environmental Lawyer Dianne Saxe argued that the CEAA 2012 "allows the federal government to create mandatory timelines for assessments of even the largest and most important projects, regardless of public opposition." (Saxe 2012)

"Now that federal environmental assessments are gone, the federal government will only assess very large, very important projects. But it's going to do them in a hurry."

On 3 August 2012 the Canadian Environmental Assessment Agency nine "designated projects" with their timelines: Enbridge Northern Gateway Pipeline Joint Review Panel (JRP) 18 months; Marathon Platinum Group Metals and Copper Mine Project (JRP): 13 months; Site C Clean Energy Project (JRP) 8.5 months; Deep Geologic Repository Project (JRP) 17 months; Enbridge Northern Gateway Project (JRP) 18 months; Jackpine Mine Expansion Project (JRP) 11.5 months; Pierre River Mine Project: 8 months; New Prosperity Gold-Copper Mine Project (JRP) 7.5 months; Frontier Oil Sands Mine Project (JRP) 8.5 months; EnCana/Cenovus Shallow Gas Infill Project (JRP) 5 months.

Saxe compares these timelines with environmental assessments for the Mackenzie Valley Pipeline. Thomas R. Berger, Royal Commissioner of the Mackenzie Valley Pipeline Inquiry (9 May 1977), worked extremely hard to ensure that industrial development on Aboriginal people's land resulted in benefits to those indigenous people.

On 22 April 2013, Official Opposition Environment critic Megan Leslie issued a statement claiming that the federal government's recent changes to "fish habitat protection, the Navigable Waters Protection Act and the Canadian Environmental Assessment Act", along with gutting existing laws and making cuts to science and research, "will be disastrous, not only for the environment, but also for Canadians' health and economic prosperity." On 26 September 2012, Leslie argued that with the changes to the Canadian Environmental Assessment Act that came into effect 6 July 2012, "seismic testing, dams, wind farms and power plants" no longer required any federal environmental assessment. She also claimed that because the CEAA 2012—which she claimed was rushed through Parliament—dismantled the CEAA 1995, the Oshawa ethanol plant project would no longer have a full federal environmental assessment. Mr. Peter Kent (Minister of the Environment) explained that the CEAA 2012 "provides for the Government of Canada and the Environmental Assessment Agency to focus on the large and most significant projects that are being proposed across the country." The 2,000 to 3,000-plus smaller screenings that were in effect under CEAA 1995 became the "responsibility of lower levels of government but are still subject to the same strict federal environmental laws." Anne Minh-Thu Quach, MP for Beauharnois—Salaberry, QC, argued that the mammoth budget bill dismantled 50 years of environmental protection without consulting Canadians about the "colossal changes they are making to environmental assessments." She claimed that the federal government is entering into "limited consultations, by invitation only, months after the damage was done."

China

The Environmental Impact Assessment Law (EIA Law) requires that an environmental impact assessment be completed prior to project construction. However, if a developer completely ignores this requirement and builds a project without submitting an environmental impact statement, the only penalty is that the environmental protection bureau (EPB) may require the developer to do a make-up environmental assessment. If the developer does not complete this make-up assessment within the designated time, only then is the EPB authorized to fine the developer. Even so, the possible fine is capped at a maximum of about US$25,000, a fraction of the overall cost of most major projects. The lack of more stringent enforcement mechanisms has resulted in a significant percentage of projects not completing legally required environmental impact assessments prior to construction.

China's State Environmental Protection Administration (SEPA) used the legislation to halt 30 projects in 2004, including three hydro-power plants under the Three Gorges Project Company. Although one month later (Note as a point of reference, that the typical EIA for a major project in the USA takes one to two years.), most of the 30 halted projects resumed their construction, reportedly having passed the environmental assessment, the fact that these key projects' construction was ever suspended was notable.

A joint investigation by SEPA and the Ministry of Land and Resources in 2004 showed that 30-40% of the mining construction projects went through the procedure of environment impact assessment as required, while in some areas only 6-7% did so. This partly explains why China has witnessed so many mining accidents in recent years.

SEPA alone cannot guarantee the full enforcement of environmental laws and regulations, observed Professor Wang Canfa, director of the centre to help environmental victims at China University of Political Science and Law. In fact, according to Wang, the rate of China's environmental laws and regulations that are actually enforced is estimated at barely 10%.

Egypt

Environmental Impact Assessment (EIA) EIA is implemented in Egypt under the umbrella of the Ministry of state for environmental affairs. The Egyptian Environmental Affairs Agency (EEAA) is responsible for the EIA services.

In June 1997, the responsibility of Egypt's first full-time Minister of State for Environmental Affairs was assigned as stated in the Presidential Decree no.275/1997. From thereon, the new ministry has focused, in close collaboration with the national and international development partners, on defining environmental policies, setting priorities and implementing initiatives within a context of sustainable development.

According to the Law 4/1994 for the Protection of the Environment, the Egyptian Environmental Affairs Agency (EEAA) was restructured with the new mandate to substitute the institution initially established in 1982. At the central level, EEAA represents the executive arm of the Ministry.

The purpose of EIA is to ensure the protection and conservation of the environment and natural resources including human health aspects against uncontrolled development. The long-term objective is to ensure a sustainable economic development that meets present needs without compromising future generations ability to meet their own needs. EIA is an important tool in the integrated environmental management approach.

EIA must be performed for new establishments or projects and for expansions or renovations of existing establishments according to the Law for the Environment.

EU

There is a wide range of instruments in the Environmental policy of the European Union. Among them the European Union has established a mix of mandatory and discretionary procedures to assess environmental impacts. European Union Directive (85/337/EEC) on Environmental Impact Assessments (known as the *EIA Directive*) was first introduced in 1985 and was amended in 1997. The directive was amended again in 2003, following EU signature of the 1998 Aarhus Convention, and once more in 2009. The initial Directive of 1985 and its three amendments have been codified in Directive 2011/92/EU of 13 December 2011. In 2001, the issue was enlarged to the assessment of plans and programmes by the so-called *Strategic Environmental Assessment (SEA) Directive* (2001/42/EC), which is now in force. Under the EU directive, an EIA must provide certain information to comply. There are seven key areas that are required:

1. Description of the project

    o Description of actual project and site description

    o Break the project down into its key components, i.e. construction, operations, decommissioning

    o For each component list all of the sources of environmental disturbance

    o For each component all the inputs and outputs must be listed, e.g., air pollution, noise, hydrology

2. Alternatives that have been considered

    o Examine alternatives that have been considered

    o Example: in a biomass power station, will the fuel be sourced locally or nationally?

3. Description of the environment

    o List of all aspects of the environment that may be affected by the development

- o Example: populations, fauna, flora, air, soil, water, humans, landscape, cultural heritage

- o This section is best carried out with the help of local experts, e.g. the RSPB in the UK

4. Description of the significant effects on the environment

- o The word significant is crucial here as the definition can vary

- o 'Significant' must be defined

- o The most frequent method used here is use of the Leopold matrix

- o The matrix is a tool used in the systematic examination of potential interactions

- o Example: in a windfarm development a significant impact may be collisions with birds

5. Mitigation

- o This is where EIA is most useful

- o Once section 4 is complete, it is obvious where impacts are greatest

- o Using this information ways to avoid negative impacts should be developed

- o Best working with the developer with this section as they know the project best

- o Using the windfarm example again construction could be out of bird nesting seasons

6. Non-technical summary (EIS)

- o The EIA is in the public domain and be used in the decision making process

- o It is important that the information is available to the public

- o This section is a summary that does not include jargon or complicated diagrams

- o It should be understood by the informed lay-person

7. Lack of know-how/technical difficulties

- o This section is to advise any areas of weakness in knowledge

- o It can be used to focus areas of future research

- o Some developers see the EIA as a starting block for poor environmental management

Annexed Projects

All projects are either classified as Annex 1 or Annex 2 projects. Those lying in Annex 1 are large scale developments such as motorways, chemical works, bridges, powerstations etc. These always require an EIA under the Environmental Impact Assessment Directive (85,337,EEC as amended). Annex 2 projects are smaller in scale than those referred to in Annex 1. Member States must determine whether these project shall be made subject to an assessment subject to a set of criteria set out in Annex 3 of codified Directive 2011/92/EU.

The Netherlands

EIA was implemented in Dutch legislation on September 1, 1987. The categories of projects that require an EIA are summarised in Dutch legislation, the Wet milieubeheer. The use of thresholds for activities makes sure that EIA is obligatory for those activities that may have considerable impacts on the environment.

For projects and plans that fit these criteria, an EIA report is required. The EIA report defines a.o. the proposed initiative, it makes clear the impact of that initiative on the environment and compares this with the impact of possible alternatives with less a negative impact.

Hong Kong

EIA in Hong Kong, since 1998, is regulated by the *Environmental Impact Assessment Ordinance 1997*.

The original proposal to construct the Lok Ma Chau Spur Line overground across the Long Valley failed to get through EIA, and the Kowloon–Canton Railway Corporation had to change its plan and build the railway underground. In April 2011, the EIA of the Hong Kong section of the Hong Kong-Zhuhai-Macau Bridge was found to have breached the ordinance, and was declared unlawful. The appeal by the government was allowed in September 2011. However, it was estimated that this EIA court case had increased the construction cost of the Hong Kong section of the bridge by HK$6.5 billion in money-of-the-day prices.

India

The Ministry of Environment, Forests and Climate Change (MoEFCC) of India has been in a great effort in Environmental Impact Assessment in India. The main laws in action are the Water Act(1974), the Indian Wildlife (Protection) Act (1972), the Air (Prevention and Control of Pollution) Act (1981) and the Environment (Protection) Act (1986),Biological Diversity Act(2002). The responsible body for this is the Central Pollution Control Board. Environmental Impact Assessment (EIA) studies need a significant amount of primary and secondary environmental data. Primary data are those collected in the field to define the status of the environment (like air quality data, water quality data etc.). Secondary data are those collected over the years that can be used to understand the existing environmental scenario of the study area. The environmental impact assessment (EIA) studies are conducted over a short period of time and therefore the understanding of the environmental trends, based on a few months of primary data, has limitations. Ideally, the primary data must be considered along with the secondary data for complete understanding of the existing environmental status of the area. In many EIA studies, the secondary data needs could be as high as 80% of the total data requirement. EIC is the repository of one stop secondary data source for environmental impact assessment in India.

The Environmental Impact Assessment (EIA) experience in India indicates that the lack of timely availability of reliable and authentic environmental data has been a major bottle neck in achieving the full benefits of EIA. The environment being a multi-disciplinary subject, a multitude of agencies are involved in collection of environmental data. However, no single organization in India tracks available data from these agencies and makes it available in one place in a form required by environmental impact assessment practitioners. Further, environmental data is not available in enhanced forms that improve the quality of the EIA. This makes it harder and more time-consuming to generate environmental impact assessments and receive timely environmental clearances from regulators. With this background, the Environmental Information Centre (EIC) has been set up to serve as a professionally managed clearing house of environmental information that can be used by MoEF, project proponents, consultants, NGOs and other stakeholders involved in the process of environmental impact assessment in India. EIC caters to the need of creating and disseminating of organized environmental data for various developmental initiatives all over the country.

EIC stores data in GIS format and makes it available to all environmental impact assessment studies and to EIA stakeholders in a cost effective and timely manner. So that we can manage that in different proportions such as remedy measures etc.,

Korea, South

Recycling culture and policy Ministry of Environment

Malaysia

In Malaysia, Section 34A, Environmental Quality Act, 1974 requires developments that have significant impact to the environment are required to conduct the Environmental impact assessment.

Nepal

In Nepal, EIA has been integrated in major development projects since the early 1980s. In the planning history of Nepal, the sixth plan (1980–85), for the first time, recognized the need for EIA with the establishment of Environmental Impact Study Project (EISP) under the Department of Soil Conservation in 1982 to develop necessary instruments for integration of EIA in infrastructure development projects. However, the government of Nepal enunciated environment conservation related policies in the seventh plan (NPC, 1985–1990). To enforce this policy and make necessary arrangements, a series of guidelines were developed, thereby incorporating the elements of environmental factors right from the project formulation stage of the development plans and projects and to avoid or minimize adverse effects on the ecological system. In addition, it has also emphasized that EIAs of industry, tourism, water resources, transportation, urbanization, agriculture, forest and other developmental projects be conducted.

In Nepal, the government's Environmental Impact Assessment Guideline of 1993 inspired the enactment of the Environment Protection Act (EPA) of 1997 and the Environment Protection Rules (EPR) of 1997 (EPA and EPR have been enforced since 24 and 26 June 1997 respectively in Nepal) to internalizing the environmental assessment system. The process institutionalized the EIA process in development proposals and enactment, which makes the integration of IEE and EIA legally binding to the prescribed projects.

New Zealand

In New Zealand, EIA is usually referred to as *Assessment of Environmental Effects* (AEE). The first use of EIA's dates back to a Cabinet minute passed in 1974 called Environmental Protection and Enhancement Procedures. This had no legal force and only related to the activities of government departments. When the Resource Management Act was passed in 1991, an EIA was required as part of a resource consent application. Section 88 of the Act specifies that the AEE must include "such detail as corresponds with the scale and significance of the effects that the activity may have on the environment". While there is no duty to consult any person when making a resource consent application, proof of consultation is almost certain required by local councils when they decide whether or not to publicly notify the consent application under Section 93.

Russian Federation

As of 2004, the state authority responsible for conducting the State EIA in Russia has been split between two Federal bodies: 1) Federal service for monitoring the use of natural resources – a part of the Russian Ministry for Natural Resources and Environment and 2) Federal Service for Ecological, Technological and Nuclear Control. The two main pieces of environmental legislation in Russia are: The Federal Law 'On Ecological Expertise, 1995 and the 'Regulations on Assessment of Impact from Intended Business and Other Activity on Environment in the Russian Federation, 2000.

Federal Service for Monitoring the use of Natural Resources

In 2006, the parliament committee on ecology in conjunction with the Ministry for Natural Resources and Environment, created a working group to prepare a number of amendments to existing legislation to cover such topics as stringent project documentation for building of potentially environmentally damaging objects as well as building of projects on the territory of protected areas. There has been some success in this area, as evidenced from abandonment of plans to construct a gas pipe-line through the only remaining habitat of the critically endangered Amur leopard in the Russian Far East.

Federal Service for Ecological, Technological and Nuclear Control

The government's decision to hand over control over several important procedures, including state EIA in the field of all types of energy projects, to the Federal Service for Ecological, Technological and Nuclear Control had caused a major controversy and criticism from environmental groups that blamed the government for giving nuclear power industry control over the state EIA.

Not surprisingly the main problem concerning State EIA in Russia is the clear differentiation of jurisdiction between the two above-mentioned Federal bodies.

Sri Lanka

The National Environmental Act, 1998 requires environmental impact assessment for large scale projects in sensitive areas. It is enforced by the Central Environmental Authority.

United States

The National Environmental Policy Act of 1969 (NEPA), enacted in 1970, established a policy of environmental impact assessment for federal agency actions, federally funded activities or federally permitted/licensed activities that in the U. S. is termed "environmental review" or simply "the NEPA process." The law also created the Council on Environmental Quality, which promulgated regulations to codify the law's requirements. Under United States environmental law an Environmental Assessment (EA) is compiled to determine the need for an *Environmental Impact Statement* (EIS). Federal or federalized actions expected to subject or be subject to significant environmental impacts will publish a Notice of Intent to Prepare an EIS as soon as significance is known. Certain actions of federal agencies must be preceded by the NEPA process. Contrary to a widespread misconception, NEPA does not prohibit the federal government or its licensees/permittees from harming the environment, nor does it specify any penalty if an environmental impact assessment turns out to be inaccurate, intentionally or otherwise. NEPA requires that plausible statements as to the prospective impacts be disclosed in advance. The purpose of NEPA process is to ensure that the decision maker is fully informed of the environmental aspects and consequences prior to making the final decision.

Environmental Assessment

An environmental assessment (EA) is an environmental analysis prepared pursuant to the National Environmental Policy Act to determine whether a federal action would significantly affect the environment and thus require a more detailed Environmental Impact Statement (EIS). The certified release of an Environmental Assessment results in either a Finding of No Significant Impact (FONSI) or an EIS.

The Council on Environmental Quality (CEQ), which oversees the administration of NEPA, issued regulations for implementing the NEPA in 1979. Eccleston reports that the NEPA regulations barely mention preparation of EAs. This is because the EA was originally intended to be a simple document used in relatively rare instances where an agency was not sure if the potential significance of an action would be sufficient to trigger preparation of an EIS. But today, because EISs are so much longer and complicated to prepare, federal agencies are going to great effort to avoid preparing EISs by using EAs, even in cases where the use of EAs may be inappropriate. The ratio of EAs that are being issued compared to EISs is about 100 to 1.

Likewise, even the preparation of an accurate EA is viewed today as an onerous burden by many entities responsible for the environmental review of a proposal. Federal agencies have responded by streamlining their regulations that implement NEPA environmental review, by defining categories of projects that by their well understood nature may be safely excluded from review under NEPA, and by drawing up lists of project types that have negligible material impact upon the environment and can thus be exempted.

Content

The Environmental Assessment is a concise public document prepared by the federal action agency that serves to:

1. briefly provide sufficient evidence and analysis for determining whether to prepare an EIS or a Finding of No Significant Impact (FONSI)

2. Demonstrate compliance with the act when no EIS is required

3. facilitate the preparation of an EIS when a FONSI cannot be demonstrated

The Environmental Assessment includes a brief discussion of the purpose and need of the proposal and of its alternatives as required by NEPA 102(2)(E), and of the human environmental impacts resulting from and occurring to the proposed actions and alternatives considered practicable, plus a listing of studies conducted and agencies and stakeholders consulted to reach these conclusions. The action agency must approve an EA before it is made available to the public. The EA is made public through notices of availability by local, state, or regional clearing houses, often triggered by the purchase of a public notice advertisement in a newspaper of general circulation in the proposed activity area.

Structure

The structure of a generic Environmental Assessment is as follows:

1. Summary

2. Introduction

   o Background

   o Purpose and Need for Action

   o Proposed Action

   o Decision Framework

   o Public Involvement

   o Issues

3. Alternatives, including the Proposed Action

   o Alternatives

   o Mitigation Common to All Alternatives

   o Comparison of Alternatives

4. Environmental Consequences

5. Consultation and Coordination

Procedure

The EA becomes a draft public document when notice of it is published, usually in a newspaper of general circulation in the area affected by the proposal. There is a 15-day review period required for an Environmental Assessment (30 days if exceptional circumstances) while the document is made available for public commentary, and a similar time for any objection to improper process. Commenting on the Draft EA is typically done in writing or email, submitted to the lead action agency as published in the notice of availability. An EA does not require a public hearing for verbal comments. Following the mandated public comment period, the lead action agency responds to any comments, and certifies either a FONSI or a Notice of Intent (NOI) to prepare an EIS in its public environmental review record. The preparation of an EIS then generates a similar but more lengthy, involved and expensive process.

Environmental Impact Statement

The adequacy of an environmental impact statement (EIS) can be challenged in federal court. Major proposed projects have been blocked because of an agency's failure to prepare an acceptable EIS. One prominent example was the Westway landfill and highway development in and along the Hudson River in New York City. Another prominent case involved the Sierra Club suing the Nevada Department of Transportation over its denial of the club's request to issue a supplemental EIS addressing air emissions of particulate matter and hazardous air pollutants in the case of widening U.S. Route 95 through Las Vegas. The case reached the United States Court of Appeals for the Ninth Circuit, which led to construction on the highway being halted until the court's final decision. The case was settled prior to the court's final decision.

Several state governments that have adopted "little NEPAs," state laws imposing EIS requirements for particular state actions. Some of those state laws such as the California Environmental Quality Act refer to the required environmental impact study as an environmental impact report.

This variety of state requirements produces voluminous data not just upon impacts of individual projects, but also in insufficiently researched scientific domains. For example, in a seemingly routine *Environmental Impact Report* for the city of Monterey, California, information came to light that led to the official federal endangered species listing of Hickman's potentilla, a rare coastal wildflower.

Transboundary Application

Environmental threats do not respect national borders. International pollution can have detrimental effects on the atmosphere, oceans, rivers, aquifers, farmland, the weather and biodiversity. Global climate change is transnational. Specific pollution threats include acid rain, radioactive contamination, debris in outer space, stratospheric ozone depletion and toxic oil spills. The Chernobyl disaster, precipitated by a nuclear accident on April 26, 1986, is a stark reminder of the devastating effects of transboundary nuclear pollution.

Environmental protection is inherently a cross-border issue and has led to the creation of transnational regulation via multilateral and bilateral treaties. The United Nations Conference on the Human Environment (UNCHE or Stockholm Conference) held in Stockholm in 1972 and the United Nations Conference on the Environment and Development (UNCED or Rio Summit, Rio Conference, or Earth Summit) held in Rio de Janeiro in 1992 were key in the creation of about 1,000 international instruments that include at least some provisions related to the environment and its protection.

The United Nations Economic Commission for Europe's Convention on Environmental Impact Assessment in a Transboundary Context was negotiated to provide an international legal framework for transboundary EIA.

However, as there is no universal legislature or administration with a comprehensive mandate, most international treaties exist parallel to one another and are further developed without the benefit of consideration being given to potential conflicts with other agreements. There is also the issue of international enforcement. This has led to duplications and failures, in part due to an inability to enforce agreements. An example is the failure of many international fisheries regimes to restrict harvesting practises. Application shall be achieved by the willing of counties authorities.

Criticism

EIA is used as a decision aiding tool rather than decision making tool. There is growing dissent about them as their influence on decisions is limited. Improved training for practitioners, guidance on best practice and continuing research have all been proposed.

EIAs have been criticized for excessively limiting their scope in space and time. No accepted procedure exists for determining such boundaries. The boundary refers to 'the spatial and temporal boundary of the proposal's effects'. This boundary is determined by the applicant and the lead assessor, but in practice, almost all EIAs address only direct and immediate on-site effects.

Development causes both direct and indirect effects. Consumption of goods and services, production, use and disposal of building materials and machinery, additional land use for activities of manufacturing and services, mining and refining, etc., all have environmental impacts. The indirect effects of development can be much higher than the direct effects examined by an EIA. Proposals such as airports or shipyards cause wide-ranging national and international effects, which should be covered in EIAs.

Broadening the scope of EIA can benefit the conservation of threatened species. Instead of concentrating on the project site, some EIAs employed a habitat-based approach that focused on much broader relationships among humans and the environment. As a result, alternatives that reduce the negative effects to the population of whole species, rather than local subpopulations, can be assessed.

Thissen and Agusdinata have argued that little attention is given to the systematic identification and assessment of uncertainties in environmental studies which is critical in situations where uncertainty cannot be easily reduced by doing more research. In line with this, Maier et al. have concluded on the need to consider uncertainty at all stages of the decision-making process. In such a way decisions can be made with confidence or known uncertainty. These proposals are justified on data that shows that environmental assessments fail to predict accurately the impacts observed. Tenney et al. and Wood et al. have reported evidence of the intrinsic uncertainty attached to EIAs predictions from a number of case studies worldwide. The gathered evidence consisted of comparisons between predictions in EIAs and the impacts measured during, or following project implementation. In explaining this trend, Tenney et al. have highlighted major causes such as project changes, modelling errors, errors in data and assumptions taken and bias introduced by people in the projects analyzed. Cardenas and Halman provide a comprehensive review on the issues of uncertainty in environmental impact assessments.

Social Impact Assessment

Social impact assessment (SIA) is a methodology to review the social effects of infrastructure projects and other development interventions. Although SIA is usually applied to planned interventions, the same techniques can be used to evaluate the social impact of unplanned events, for example disasters, demographic change and epidemics.

Definition

The origins of SIA largely derive from the environmental impact assessment (EIA) model, which first emerged in the 1970s in the U.S, as a way to assess the impacts on society of certain development schemes and projects before they go ahead - for example, new roads, industrial facilities, mines, dams, ports, airports, and other infrastructure projects. In the United States under the National Environmental Policy Act, social impact assessments are federally mandated and performed in conjunction with environmental impact assessments. SIA has been incorporated into the formal planning and approval processes in several countries, in order to categorize and assess how major developments may affect populations, groups, and settlements. Though the social impact assessment has long been considered subordinate to the environmental impact assessment, new models, such as the Environmental Social Impact Assessment (ESIA), take a more integrated approach where equal weight is given to both the social and environmental impact assessments.

Contents

Definitions for "social impact assessment" vary by different sectors and applications. According to the International Association for Impact Assessment, "Social impact assessment includes the processes of analyzing, monitoring and managing the intended and unintended social consequences, both positive and negative, of planned interventions (policies, programs, plans, projects) and any social change processes invoked by those interventions. Its primary purpose is to bring about a more sustainable and equitable biophysical and human environment."

SIA overlaps with monitoring and evaluation (M&E). Evaluation is particularly important in the areas of:

1. public policy,

2. health and education initiatives, and

3. international development projects more generally, whether conducted by governments, international donors, or NGOs.

In all these sectors, there is a case for conducting SIA and evaluations at different stages.

The Hydropower Sustainability Assessment Protocol is a sector specific method for checking the quality of environmental and social assessments and management plans.

Non-experts and local people should participate in the design and implementation of proposed developments or programmes. This can be achieved in the process of doing an SIA, through adopting a participatory and democratic research process. Some SIAs go further than this, to adopt an advocacy role. For example, several SIAs carried out in Queensland, Australia, have been conducted by consultants working for local Aboriginal communities who oppose new mining projects on ancestral land. A rigorous SIA report, showing real consequences of the projects and suggesting ways to mitigate these impacts, gives credibility and provides evidence to take these campaigns to the planning officers or to the courts.

Landscape Assessment

Landscape assessment is a sub-category of environmental impact assessment (EIA) concerned with quality assessment of the landscape. Landscape quality is assessed either as part of a strategic planning process or in connection with a specific development which will affect the landscape. These methods are sub-divided into area-based assessments or proposal-driven assessments, respectively. The term 'landscape assessment' can be used to mean either visual assessment or character assessment. Since landscape assessments are intended to help with the conservation and enhancement of environmental goods, it is usually necessary to have a fully geographical landscape assessment as a stage in the process of EIA and landscape planning. During the initial phases of a project, such as site selection and design concept, the landscape architect begins to identify areas of opportunity or setbacks that may provide constraints. The architect prepares alternative options in order to compare their assessments and identifies the proposal's which allow for the least adverse effects on the landscape or views. A landscape professional works with a design team to review potential effects as the team develops a sustainable proposal. Upon developing a design proposal, the landscape professional will identify and describe the landscape and visual effects that may occur and suggest mitigation measures to be taken in order to reduce negative effects and maximize benefits, if any.

Landscape and Visual Impact Assessment (LVIA)

This process, which operates within the larger framework of Environmental Impact Assessment, strives to ensure that any of the effects of change are taken into account in the decision-making process of a project. It is essential that any possible change or development to the landscape or views around a project be evaluated throughout the planning and design phase of a project. Thus, landscape assessment is sub-divided into two types: visual assessment and character assessment.

Visual Assessment

This would look at how changes in the landscape could alter the nature and extent of visual effects and qualities relating to locations and proposals and how it effects specific individuals or groups of people. UK-related guidance on the preparation of these assessments is given in the 3rd edition of the *Guidelines for Landscape and Visual Impact Assessment* published by Routledge on behalf of the Landscape Institute & Institute of Environmental Management, 2013.

Character Assessment

This includes assessments of each aspect of the landscape: geology, hydrology, soils, ecology, settlement patterns, cultural history, scenic characteristics, land use etc. It refers to the assessment of the individual components of a landscape listed above when experiencing change. It typically includes distinct descriptive and evaluative components. Guidance on the preparation of these assessments is given in *Landscape Character assessment: Guidance for England and Scotland* published by the Countryside Agency & Scottish Natural Heritage, April 2002.

Area-based Assessment

This assessment can be completed at the regional scale as well as district, city, or catchment scale. This process is used to determine where landscape management is required. The process consists of three stages: landscape description, landscape characterization, and landscape evaluation.

Landscape Description

The first step in completing an area-based assessment is to compile data in order to identify components of the landscape within a project area. Components of a landscape range from landform, geology, soil, vegetation cover, drainage patterns, built development, land uses, infrastructure, and heritage sites to cultural meaning. This step in the assessment of a landscape is not site specific, but instead, a general description of the landscape.

Landscape Characterization

This step in the assessment refers to the process of identification, mapping, and description of landscape character areas. Landscape character areas are geographical areas made up by a combination of individual landscape components that make one area different from another area. The characterization of a landscape should begin to define the boundaries of the area being assessed.

Landscape Evaluation

The last step in an area-based assessment is the evaluation process. This is a critical phase in the assessment process because landscape evaluation's are the driving force behind the planning and design of landscapes. Here, the assessment should identify important landscapes or natural features and assign rankings and priorities to features that require management.

Drawback

The evaluation process in the assessment is subjective and dependent on the professional completing the assessment and thus creates a controversial issue. Therefore, the assessment should always be completed by professionals who are trained to make accurate judgments of a landscape.

Proposal-driven Assessment

Landscape quality can be assessed in connection with a specific development which will affect the landscape. Such assessment requires that a professional submit a development proposal. This approach to complete an assessment serves to identify the potential effects on landscape values brought forth by a certain proposal. The specific proposal is analyzed to evaluate the effects it may have on the landscape or character of the landscape, as well as the proposal's effect on the composition of available views. In a proposal-driven assessment, the area involved should include the site of the project as well as its immediate surroundings. This assessment should produce a detailed description of any physical changes to the landscape as well as a description and analysis of the effect these changes will have. This process should evaluate the importance of character, landscape, and visual amenity. Ultimately, this approach is effective if, and only when measures that can mitigate the effect of a given development proposal are identified.

Benefits of Environmental Assessment

Most governments and donor agencies acknowledge the contribution of EA to improved project design. The weakness of EA in the past has been largely due to poor techniques and the failure to pay attention to findings at the implementation stage (ESSA Technologies 1994). A review of current environmental practices found the major benefits of the EA process for project sponsors to be:

- Reduced cost and time of project implementation.

- Cost-saving modifications in project design.

- Increased project acceptance.

- Avoided impacts and violations of laws and regulations.

- Improved project performance.

- Avoided treatment/clean up costs.

The benefits to local communities from taking part in environmental assessments include:

- A healthier local environment (forests, water sources, agricultural potential, recreational potential, aesthetic values, and clean living in urban areas).

- Improved human health.

- Maintenance of biodiversity.

- Decreased resource use.

- Fewer conflicts over natural resource use.

- Increased community skills, knowledge and pride

Principle of EIA

It is important to recognise that there is a general principle of assessment that applies to EIA, and to other assessment processes. There are several other processes that relate closely to the review of environmental impacts that may result from a proposed project. The following are well recognised processes:

- Social Impact Assessment

- Risk Assessment

- Life Cycle Analysis

- Energy Analysis

- Health Impact Assessment

- Regulatory Impact Assessment

- Species Impact Assessment

- Technology Assessment

- Economic Assessment

- Cumulative Impact Assessment

- Strategic Environmental Assessment

- Integrated Impact Assessment

Some, like Energy Analysis, focus on a particular part of the environment. Others, like Life Cycle Analysis, enable the consideration of all those parts of the environment that are relevant to the assessment. Also, depending on how the terms, like health, are defined for the study you may find that it is covering most of the issues that would be found in an EIA. For example a Technology Assessment does include review of the impacts on ecosystems, air quality and the like. Similarly, if the definition of environmental is taken broadly for an EIA, then the EIA may cover the issues of the other assessment processes; for example:

- Social aspects (such as impacts on employment, community interaction);

- Risks (such as threats to native animals, water supplies);

- Life cycle (such as the impacts at each stage of the project design through to operation and closure); and

- Energy (such as use of non-renewable energy sources, Greenhouse gas emissions), etc

So there is the potential for a lot of connections between the different forms of assessment. The essential difference between them is how the terms, or scope of assessment, are defined narrowly, or broadly. Otherwise they all follow the same general principle.

With all the assessment approaches noted above, they are designed to identify potential impacts of a development, action or project. To do this the assessor needs to use personal experience and the experiences of others (including available knowledge) to think broadly about the changes that are possible, and whether those impacts will be positive or negative.

Particular approaches emphasis specific types of impacts (i.e. on health, on social groups). All have basically the same approach, although each may have its own individual language and detailed techniques. Most of the assessment processes also include a second step. After identifying the impacts, they also consider what may be needed to avoid or reduce adverse impacts.

Purposes of EIA

EIA is a process with several important purposes, which can be categorised as follows:

- To facilitate decision-making: For the decision-maker, for example the local authority, it provides a systematic examination of the environmental implications of a proposed action, and sometimes alternatives, before a decision is taken. The decision-maker along with other documentation relating to the planned activity can consider the environment impact statement (EIS).

- To aid in the formation of development: Many developers see EIA as another set of hurdles for them to cross in order to proceed with their various activities. They may also see the process involved in obtaining the permission from various authorities as costly and time-consuming. In reality, however, EIA can be of great benefit to them, since it can provide a framework for considering location and design issues and environmental issues in parallel. It can be an aid to the formulation of developmental actions, indicating areas where the project can be modified to minimise or eliminate altogether the adverse impacts on the environment. The consideration of environmental impacts early in the planning life of a development can lead to environmentally sensitive development; to improved relations between the developer, the planning authority and the local communities; to a smoother planning permission process and sometimes to a worthwhile financial return on the expenditure incurred.

- To be an instrument for sustainable development: The key characteristics of sustainable development include maintaining the overall quality of life, maintaining continuing access to natural resources and avoiding lasting environmental damage. Institutional responses to sustainable development are,

therefore, required at several levels. For example, issues of global concern, such as ozone-layer depletion, climate change, deforestation and biodiversity loss, require a global political commitment to action. The United Nations Conference on Environment and Development (UNCED) held in Rio de Janeiro in 1992 was an example of international concern and also of the problems of securing concerted action to deal with such issues. Governments have recognised the interaction of economic and social development and the ecosystems, and the reciprocal impact between human actions and the biogeophysical world. While there are attempts to manage this interaction better, investigation reveal disquieting trends that could have devastating consequences for the quality of the environment. These trends are likely to be more pronounced in developing countries where, because of greater rates of population growth and lower current living standards, there is more pressure on environmental resources.

In short, an interaction among the resources, sectors and policies is necessary for sustainable development as illustrated in Figure below, and EIA contributes to this process:

Sustainable Development: An Illustration

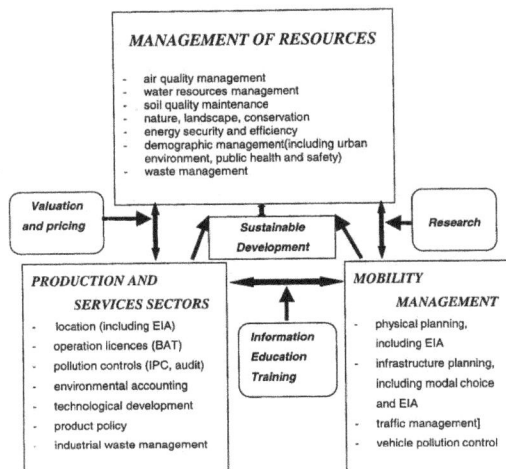

Steps in EIA Process

EIA represents a systematic process that examines the environmental consequences of the development actions, in advance. The emphasis of a EIA is on prevention and, therefore, is more proactive than reactive in nature. The EIA process involves a number of steps, some of which are listed below:

- Project screening: This entails the application of EIA to those projects that may have significant environmental impacts. It is quite likely, however, that screening is done partly by the EIA regulations, operating in a country at the time of assessment.

- Scoping: This step seeks to identify, at an early stage, the key, significant environmental issues from among a host of possible impacts of a project and all the available alternatives.

- Consideration of alternatives: This seeks to ensure that the proponent has considered other feasible approaches, including alternative project locations, scales, processes, layouts, operating condition and the no-action option.

- Description of the project/development action: This step seeks to clarify the purpose and rationale of the project and understand its various characteristics, including the stages of development, location and processes.

- Description of the environmental baseline: This includes the establishment of both the present and future state of the environment, in the absence of the project, taking into account the changes resulting from natural events and from other human activities.

- Identification of key impacts: This brings together the previous steps with a view to ensuring that all potentially significant environmental impacts (adverse and beneficial) are identified and taken into account in the process.

- The prediction of impacts: This step aims to identify the likely magnitude of the change (i.e., impact) in the environment when the project is implemented in comparison with the situation when the project is not carried out.

- Evaluation and assessment of significance: This seeks to assess the relative significance of the predicted impacts to allow a focus on key adverse impacts. Formal definition of significance is the product of consequence and likelihood as Significance =consequence X Likelihood

- Mitigation: This involves the introduction of measures to avoid, reduce, remedy or compensate for any significant adverse impacts.

- Public consultation and participation: This aims to assure the quality, comprehensiveness and effectiveness of the EIA, as well as to ensure that the public's views are adequately taken into consideration in the decision-making process.

- EIS presentation: This is a vital step in the process. If done badly, much good work in the EIA may be negated.

- Review: This involves a systematic appraisal of the quality of the EIS, as a contribution to the decision-making process.

- Decision-making: At this stage, decisions are made by the relevant authority of the EIS (including consultation responses) together with other material considerations as to whether to accept, defer or reject the project.

- Post-decision monitoring: This involves the recording of outcomes associated with development impacts, after the decision to proceed with the project. It can contribute to effective project management.

- Auditing: This follows monitoring and involves comparing actual outcomes with predicted outcomes, and can be used to assess the quality of predictions and the effectiveness of mitigation. It provides a vital step in the EIA learning process.

The following figure illustrates the steps involved in the EIA process:

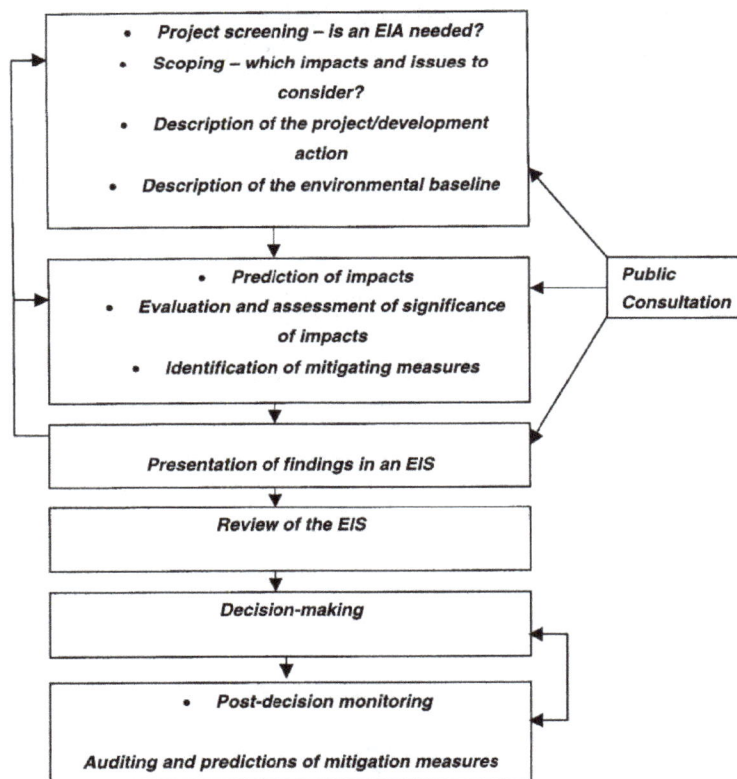

Note that the actual EIA process is not so linear and sequential as it seems to suggest. In other words, it is a cyclical process involving feedback and interaction among the various steps and the sequence of the steps may also vary.

Hierarchy in EIA

The EIA studies are broadly categorised as:

(i) Site selection studies: These studies involve an evaluation of the alternative sites with respect to environmental and project attributes such as proximity to raw materials, infrastructure facilities, markets, etc. These studies aim at ranking site alternatives for objective decision-making.

(ii) Rapid or comprehensive studies: Rapid studies refer to the assessment based on a one-season monitoring (i.e., 3- month period), whereas comprehensive studies relate to the assessment based on a three-seasons monitoring (i.e., 9- month period) of baseline data. Rapid EIA facilitates decision-making in situations where a fair amount of knowledge exists about the proposed site or the impacts of the proposed development. It also helps in identifying significant issues for comprehensive EIA. Essentially, rapid and comprehensive studies differ with respect to timeframes required for baseline data collection.

(iii) Regional studies: These relate to the development in/of a region based on seasonal data collection and address themselves to the analysis of assimilative capacity of air, water and land components of the environment.

(iv) Carrying capacity studies: The scope of a carrying capacity study is extended to the analysis of supportive capacity in the region with respect to resource availability/ utilisation, supply/demand, infrastructure/congestion and assimilative capacity/residuals. Carrying capacity has been discussed in detail in Unit 9.

n the last two decades, national governments and also financial institutions have realised that EIA has to be an integral part of the project life cycle: from project conceptualisation to post implementation corrective action.

EIA Cycle

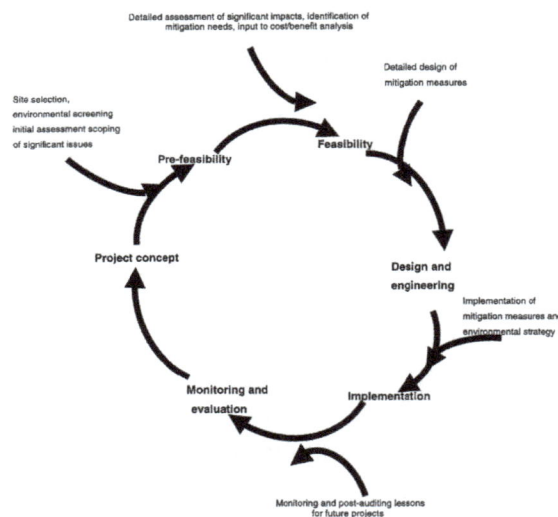

Environmental Impact Statement

An environmental impact statement (EIS), under United States environmental law, is a document required by the National Environmental Policy Act (NEPA) for certain actions "significantly affecting the quality of the human environment". An EIS is a tool for decision making. It describes the positive and negative environmental effects of a proposed action, and it usually also lists one or more alternative actions that may be chosen instead of the action described in the EIS. Several U.S.

state governments require that a document similar to an EIS be submitted to the state for certain actions. For example, in California, an Environmental Impact Report (EIR) must be submitted to the state for certain actions, as described in the California Environmental Quality Act (CEQA). One of the primary authors of the act is Lynton K. Caldwell.

Purpose

The purpose of the NEPA is to promote informed decision-making by federal agencies by making "detailed information concerning significant environmental impacts" available to both agency leaders and the public. The NEPA was the first piece of legislation that created a comprehensive method to assess potential and existing environmental risks at once. It also encourages communication and cooperation between all the actors involved in environmental decisions, including government officials, private businesses, and citizens.

In particular, an EIS acts as an enforcement mechanism to ensure that the federal government adheres to the goals and policies outlined in the NEPA. An EIS should be created in a timely manner as soon as the agency is planning development or is presented with a proposal for development. The statement should use an interdisciplinary approach so that it accurately assesses both the physical and social impacts of the proposed development. In many instances an action may be deemed subject to NEPA's EIS requirement even though the action is not specifically sponsored by a federal agency. Eccleston identifies instances that may 'federalize' such actions for the purposes of NEPA. These factors may include actions that receive federal funding, federal licensing or authorization, or that are subject to federal control.

Not all federal actions require a full EIS. If the action may or may not cause a significant impact the agency can first prepare a smaller, shorter document called an Environmental Assessment (EA). The finding of the EA determines whether an EIS is required. If the EA indicates that no significant impact is likely, then the agency can release a finding of no significant impact (FONSI) and carry on with the proposed action. Otherwise, the agency must then conduct a full-scale EIS. Most EAs result in a FONSI. A limited number of federal actions may avoid the EA and EIS requirements under NEPA if they meet the criteria for a categorical exclusion (CATEX). A CATEX is usually permitted when a course of action is identical or very similar to a past course of action and the impacts on the environment from the previous action can be assumed for the proposed action, or for building a structure within the footprint of an existing, larger facility or complex. For example, two proposed sections of Interstate 69 in Kentucky were granted a CATEX from NEPA requirements as these portions of I-69 will be routed over existing freeways requiring little more than minor spot improvements and a change of highway signage. Additionally, a CATEX can be issued during an emergency when time does not permit the preparation of an EA or EIS. An example of the latter is when the Federal Highway Administration issued a CATEX to construct the replacement bridge in the wake of the I-35W Mississippi River Bridge Collapse.

Contrary to a widespread misconception, NEPA does not prohibit the federal government or its licensees/permittees from harming the environment, but merely requires that the prospective impacts be understood and disclosed in advance. The intent of NEPA is to help key decisionmakers and stakeholders balance the need to implement an action with its impacts on the surrounding human and natural environment, and provide opportunities for mitigating those impacts while keeping the cost and schedule for implementing the action under control. However, many activities require various federal permits to comply with other environmental legislation, such as the Clean Air Act, the Clean Water Act, Endangered Species Act and Section 4(f) of the Federal Highway Act to name a few. Similarly, many states and local jurisdictions have enacted environmental laws and ordinances, requiring additional state and local permits before the action can proceed. Obtaining these permits typically requires the lead agency to implement the Least Environmentally Damaging Practicable Alternative (LEDPA) to comply with federal, state, and local environmental laws that are ancillary to NEPA. In some instances, the result of NEPA analysis leads to abandonment or cancellation of the proposed action, particularly when the "No Action" alternative ends up being the LEDPA.

Layout

An EIS typically has four sections:

- An Introduction including a statement of the Purpose and Need of the Proposed Action.

- A description of the Affected Environment.

- A Range of Alternatives to the proposed action. Alternatives are considered the "heart" of the EIS.

- An analysis of the environmental impacts of each of the possible alternatives. This section covers topics such as:

- Impacts to threatened or endangered species

- Air and water quality impacts

- Impacts to historic and cultural sites, particularly sites of significant importance to Indigenous peoples.

- Social and Economic impacts to local communities, often including consideration of attributes such as impacts to available housing stock, economic impacts to businesses, property values, aesthetics and noise within the affected area

- Cost analysis for each alternative, including costs to mitigate expected impacts, to determine if the proposed action is a prudent use of taxpayer dollars

While not required in the EIS, the following subjects may be included as part of the EIS or as separate documents based on agency policy.

- Financial Plan for the proposed action identifying the sources of secured funding for the action. For example, the Federal Highway Administration has started requiring states to include a financial plan showing that funding has been secured for major highway projects before it will approve an EIS and issue a Record of Decision.

- An Environmental Mitigation Plan is often requested by the Environmental Protection Agency (EPA) if substantial environmental impacts are expected from the preferred alternative.

- Additional documentation to comply with state and local environmental policy laws and secure required federal, state, and local permits before the action can proceed.

Every EIS is required to analyze a No Action Alternative, in addition to the range of alternatives presented for study. The No Action Alternative identifies the expected environmental impacts in the future if existing conditions were left as is with no action taken by the lead agency. Analysis of the No Action Alternative is used to establish a baseline upon which to compare the proposed "Action" alternatives.

NEPA Process

The NEPA process is designed to involve the public and gather the best available information in a single place so that decision makers can be fully informed when they make their choices.

The process has the following steps:

- Proposal: In this stage, the needs and objectives of a project have been decided, but the project has not been financed.

- Categorical Exclusion (CATEX): As discussed above, the government may exempt an agency from the process. The agency can then proceed with the project and skip the remaining steps.

- Environmental Assessment (EA): The proposal is analyzed in addition to the local environment with the aim to reduce the negative impacts of the development on the area.

- Finding of No Significant Impact (FONSI): Occurs when no significant impacts are identified in an EA. A FONSI typically allows the lead agency to proceed without having to complete an EIS.

Environmental Impact Statement

- Scoping: The first meetings are held to discuss existing laws, the available information, and the research needed. The tasks are divided up and a lead group is selected. Decision makers and all those involved with the project can attend the meetings.

- Notice: The public is notified that the agency is preparing an EIS. The agency also provides the public with information regarding how they can become involved in the process. The agency announces its project proposal with a notice in the Federal Register, notices in local media, and letters to citizens and groups that it knows are likely to be interested. Citizens and groups are welcome to send in comments helping the agency identify the issues it must address in the EIS (or EA).

- Draft EIS (DEIS): Based on both agency expertise and issues raised by the public, the agency prepares a Draft EIS with a full description of the affected environment, a reasonable range of alternatives, and an analysis of the impacts of each alternative.

- Comment: Affected individuals then have the opportunity to provide feedback through written and public hearing statements.

- Final EIS (FEIS) and Proposed Action: Based on the comments on the Draft EIS, the agency writes a Final EIS, and announces its Proposed Action. The public is not invited to comment on this, but if they are still unhappy, or feel that the agency has missed a major issue, they may protest the EIS to the Director of the agency. The Director may either ask the agency to revise the EIS, or explain to the protester why their complaints are not actually taken care of.

- Re-evaluation: Prepared following an approved FEIS or ROD when unforeseen changes to the proposed action or its impacts occurs, or when a substantial period of time has passed between approval of an action and the planned start of said action. Based on the significance of the changes, three outcomes may result from a re-evaluation report: (1) the action may proceed with no substantive changes to the FEIS, (2) significant impacts are expected with the change that can be adequately addressed in a Supplemental EIS (SEIS), or (3) the circumstances force a complete change in the nature and scope of the proposed action, thereby voiding the pre-existing FEIS (and ROD, if applicable), requiring the lead agency to restart the NEPA process and prepare a new EIS to encompass the changes.

- Supplemental EIS (SEIS): Typically prepared after either a Final EIS or Record of Decision has been issued and new environmental impacts that were not considered in the original EIS are discovered, requiring the lead agency to re-evaluate its initial decision and consider new alternatives to avoid or mitigate the new impacts. Supplemental EISs are also prepared when the size and scope of a federal action changes, when a significant period of time has lapsed since the FEIS was completed to account for changes in the surrounding environment during that time, or when all of the proposed alternatives in an EIS are deemed to have unacceptable environmental impacts and new alternatives are proposed.

- Record of Decision (ROD): Once all the protests are resolved the agency issues a Record of Decision which is its final action prior to implementation. If members of the public are still dissatisfied with the outcome, they may sue the agency in Federal court.

Often, the agencies responsible for preparing an EA or EIS do not compile the document directly, but outsource this work to private-sector consulting firms with expertise in the proposed action and its anticipated effects on the environment. Because of the intense level of detail required in analyzing the alternatives presented in an EIS or EA, such documents may take years or even decades to compile, and often compose of multiple volumes that can be thousands to tens of thousands of pages in length.

To avoid potential conflicts in securing required permits and approvals after the ROD is issued, the lead agency will often coordinate with stakeholders at all levels, and resolve any conflicts to the greatest extent possible during the EIS process. Proceeding in this fashion helps avoid interagency conflicts and potential lawsuits after the lead agency reaches its decision.

Tiering

On exceptionally large projects, especially proposed highway and railroad corridors that cross long distances, the lead agency may use a two-tiered process prior to implemeting the proposed action. In such cases, the Tier I EIS would analyze the potential socio-environmental impacts along a general corridor, but would not identify the exact location of where the action would occur. A Tier I ROD would be issued approving the general area where the action would be implemented. Following the Tier I ROD, the approved Tier I area is further broken down into subareas, and a Tier II EIS is then prepared for each subarea, that identifies the exact location of where the proposed action will take place. The preparation of Tier II EISs for each subarea proceeds at its own pace, independent from the other subareas within the Tier I area. For example, parts of the proposed Interstate 69 extension in Indiana and Texas are being studied through a two-tiered process.

Strengths

By requiring agencies to complete an EIS, the act encourages them to consider the environmental costs of a project and introduces new information into the decision-making process. The NEPA has increased the influence of environmental analysts and agencies in the federal government by increasing their involvement in the development process. Because an EIS requires expert skill and knowledge, agencies must hire environmental analysts. Unlike agencies who may have other priorities, analysts are often sympathetic to environmental issues. In addition, this feature introduces scientific procedures into the political process.

Limitations

The differences that exist between science and politics limit the accuracy of an EIS. Although analysts are members of the scientific community, they are affected by the political atmosphere. Analysts do not have the luxury of an unlimited time for research. They are also affected by the different motives behind the research of the EIS and by different perspectives of what constitutes a good analysis. In addition, government officials do not want to reveal an environmental problem from within their own agency.

Citizens often misunderstand the environmental assessment process. The public does not realize that the process is only meant to gather information relevant to the decision. Even if the statement predicts negative impacts of the project, decision makers can still proceed with the proposal.

Environmental impact statements presented to citizens and government officials frequently include very precise data. However, the quality and context of the data, such as the margin of error and the range, is omitted.

The environmental impact statement (EIS) provides documentation of the information and estimates derived from the various steps in the EIA process. The information contained in a EIS provides the decision-makers/regulators with valuable information that could ultimately contribute to either the abandonment or substantial modification of a proposed development action. A typical EIS contains the following three parts:

- Part 1 – Methods and key issues: This part deals with the statement of methods used and a summary of key issues.

- Part 2 – Background to the proposed development: This part deals with preliminary studies (i.e., need, planning, alternatives, site selection, etc.), site description/baseline conditions, description of proposed development and construction activities and programmes.

- Part 3 – Environmental impact assessments on topic areas: This part deals with land use, landscape and visual quality, geology, topography and soils, hydrology and water quality, air quality and climate, terrestrial and aquatic ecology, noise, transport, socio-economic and interrelationships between effects.

Impact Indicators

An impact indicator is an element or a parameter that provides a measure (in at least some qualitative sense) of the significance of the effect, i.e., the magnitude of an environmental impact. Some indicators such as morbidity and mortality statistics and crop yields have associated numerical scales. Other impact indicators, however, can only be ranked as 'good', 'better', 'best' or 'acceptable', 'unacceptable', etc. The selection of a set of indicators is often a crucial step in the impact assessment process, requiring input from the decision-maker. In the absence of relevant goals or policies, the assessor himself or herself may suggest some indicators and scales, but he or she should not proceed with the assessment until his or her proposals are accepted.

The most widely used impact indicators are those within statutory laws, acts, i.e., indicators such as air and water quality standards that have statutory authority. For example, the problem of designing an environmentally acceptable oil-fired generating station is simplified for the engineers, if they are given one or both of the following:

- Emission standards for various pollutants.

- Air and water quality standards.

These standards integrate the worth that a jurisdiction places on clean air and clear water. The numerical values that have been derived from examination of the available toxicological matter are data relating polluting dosages to health and vegetation effects, combined with a consideration of the best practical technology. Factors such as the displacement of arable land by industry are also equally important. A EIA that ignores these other components is incomplete and sometimes misleading (Munn, 1979).

Evolution of Eia

To understand the use of EIA as a tool for environmental management, let us discuss how EIA has evolved over the years.

Evolution of EIA Worldwide

United States of America was the first country to assign mandatory status to EIA through its National Environmental Protection Act (NEPA) of 1969. A host of industrialised countries have since implemented EIA procedures. Canada, Australia, the Netherlands and Japan adopted EIA legislation in 1973, 1974, 1981 and 1984, respectively. In July 1985, the European Community (EC) issued a directive making environmental assessments mandatory for certain categories of projects (Wood, 1994).

Among the developing countries, Columbia was the first Latin American country to institute a system of EIA in 1974. In Asia and the Pacific region, Thailand and the Philippines have long established procedures for EIA. EIA was made mandatory in Sri Lanka in 1984. The EIA process in Africa is sketchy, although a number of nations including Rwanda, Botswana and Sudan have some experience of EIA (Wathern, 1988).

Bilateral and multilateral agencies have also recognised the value of EIA as a decision-making tool. The Organisation for Economic Co-Operation and Development (OECD) issued recommendations on EIA to its constituent States in 1974 and 1979, and for development aid projects in 1986. OECD issued guidelines for good practices in EIA in 1992 (OECD, 1992). United Nations Environment Programme (UNEP) in 1980 provided guidance on EIA of the development proposals (UNEP, 1980) and supported research on EIA in developing countries (Ahmad and Swamy, 1985). UNEP, in 1987, set out goals and principles of EIA for the member countries and provided guidance on basic procedures for EIA in 1988.

The World Conservation Strategy pinpointed the need to integrate environmental considerations with development in 1980 (IUCN, 1980). EIA became an integral part of World Bank policy in 1987 which states that environmental issues must be addressed as part of overall economic policy. In 1989, the World Bank issued the Operational Directive on Environmental Assessment (O.D. 4.00), which was revised and updated in October 1991 (O.D. 4.01). Asian Development Bank in 1990 published guidelines for EIA (ADB, 1990). Importance of EIA was echoed in the Brundtland Report (WCED, 1987), and at United Nations Earth Summit on environment and development held at Rio de Janeiro in 1992 (UNCED, 1992). As foreseen by Garner and O'Riodan (1982) development of EIA, as a tool for decision-making world-over, has emerged through the following stages:

- No formal accounting, decisions made on interest group lobbying and engineering feasibility; primary emphasis on economic development.

- Conventional cost-benefit analysis; emphasis on efficiency criterion and engineering feasibility; major concern still on economic development.

- Innovative cost-benefit analysis, use of multiple objectives and discount rates, imaginative proxy pricing mechanisms; economic development as one of the objectives.

- EIA mainly concerned with describing the repercussions of the proposals on bio-physical processes; economic development still primary objective.

- EIA with more attention paid to socio-cultural as well as biophysical systems, economic development but not the sole objective.

The summary of evolution of EIA in various countries is presented in the table:

Evolution of EIA Worldwide

Australia	Environmental Protection (Impact of Proposals) Act 1974, Commonwealth of Australia
Bangladesh	No specific EIA legislation, however there was a Declaration that Environmental Impact Assessments should be carried out for all major development projects, 1995
China	Environmental Protection Law, 1979
USA (California)	California Environmental Quality Act (CEQA) of 1971
Canada	Federal Environmental Assessment and Review Process Guidelines Order 1984, Canada

France	Law on Protection de la Nature, 1978
India	Notifications dated May 5, 1994 under the Environment Protection Act, 1986
Japan	Principles for Implementing EIA by Environmental Agency, 1984
Malaysia	Environmental Quality (Prescribed Activity) (EIA) Order, 1987
New Zealand	Resource Management Act 1991, New Zealand
Philippines	Presidential Decree (PD) 1151 Philippines Environment Policy, 1975 PD 1586 Establishing the Environmental Impact Statement (EIS), 1978 Rules and Regulations to Implement the EIS System, 1987
Sri Lanka	National Environmental Act 1980, amended in 1986
Thailand	Improvement and Conservation of National Environmental Quality Act 1975, amended in 1978
The Netherlands	EIA Policy, 1986
United States	US Environmental Policy Act, 1969
Vietnam	Environmental Protection Law, 1994
Western Australia	Environmental Protection Act 1986
West Germany	Cabinet Resolution, 1975

Evolution of EIA in India

EIA in India was started in 1976-77, when the Planning Commission asked the then Department of Science and Technology to examine the river-valley projects from the environmental angle. This was subsequently extended to cover those projects, which required approval of the Public Investment Board. These were administrative decisions, and lacked the legislative support. The Government of India enacted the Environment (Protection) Act on 23rd May 1986. To achieve the objectives of the Act, one of the decisions taken was to make EIA statutory. After following the legal procedure, a notification was issued on 27th January 1994 and subsequently amended on 4th May 1994, 10th April 1997 and 27th January 2000 making environmental impact assessment statutory for 30 activities. This is the principal piece of legislation governing EIA in India. Besides this, the Government of India under Environment (Protection) Act 1986 issued a number of notifications, which are related to environmental impact assessment. These are limited to specific geographical areas, and are summarised below:

- Prohibiting location of industries except those related to Tourism in a belt of 1 km from high tide mark from the Revdanda Creek up to Devgarh Point (near Shrivardhan) as well as in 1 km belt along the banks of Rajpuri Creek in Murud Janjira area in the Raigarh district of Maharashtra (6th January 1989).

- Restricting location of industries, mining operations and regulating other activities in Doon Valley (1st February 1989).

- Regulating activities in the coastal stretches of the country by classifying them as coastal regulation zone and prohibiting certain activities (19th February 1991).

- Restricting location of industries and regulating other activities in Dahanu Taluka in Maharashtra (6th June 91).

- Restricting certain activities in specified areas of Aravalli Range in the Gurgaon district of Haryana and Alwar district of Rajasthan (7th May 1992).

- Restricting industrial and other activities, which could lead to pollution and congestion in the north west of Numaligarh in Assam (July 1996).

Forecasting Environmental Changes

A EIA should be able to, among others, predict the nature and extent of the impact of human activities on the environment. The Table below gives a list of human-induced environmental changes, which can be either benign or malignant to the environment:

Environmental Changes	
Medium	**Changes and Rates of Change in**
Soil	Quality (e.g., depth, structure, fertility, degree of stalinisation or acidification, etc.)
	Stability
	Area of arable land
Air	Quality
	The climatic elements
Water	Quantity
	Quality

Environmental Changes	
Medium	**Changes and Rates of Change in**
	Season ability
	Area of human-made lakes Extent of irrigation canals
Biota	Abundance/scarcity of species or genetic
	resources
	Extent of crops, ecosystems, vegetation and forests
	Diversity of species
	Extent of provision of nesting grounds, etc., for migratory species
	Abundance/scarcity of pests and disease organisms.

Of importance here are not only estimates of changes in environmental quality but also estimates of rate of change. A slow change may be acceptable, especially if it leads to a new stability, whereas rapid change or large fluctuations may place intolerable burdens on ecosystems. Of equal or perhaps greater importance is the degree of irreversibility of an environmental change, which will be either absolute, as in the extinction of a species, or partly absolute in that the situation can only be reversed over long periods of time or with unacceptable expenditures of money and energy, as in the case of catastrophic erosion.

A typical EIA contains information on the following three areas, as they relate to environmental effects:

(i) A determination of the initial reference state.

(ii) An estimate of the future state without action.

(iii) An estimate of the future state with action.

We will describe each of these, next.

Establishment of the initial Reference State

An assessment of environmental change pre-supposes knowledge about the present state. It will be necessary, therefore, to select attributes that may be used to estimate this state. Some of these will be directly measurable; others will only be capable of being recorded within a series of defined categories, or ranked in ascending or descending order of approximate magnitude. Difficult decisions need to be made about the population (i.e., in a statistical sense), which is to be represented by the measured variables, and the extent to which the sub-division of this population into geographical regions, ecosystems, etc., is either feasible or necessary. In fact, it must be emphasised that the establishment of an initial reference state is difficult because not only are environmental systems dynamic but also they contain cyclical and random components.

Predicting the Future State in the Absence of Action

n order to provide a fair basis for examining the impact of human activities on the environment, a EIA must estimate the future environmental states in the absence of action. As an example, the population of a species of animal or fish may already be declining, due to over-grazing or over-fishing, even before a smelter is built. This part of analysis is largely a scientific problem, requiring skills drawn from many disciplines. The prediction will often be uncertain but the degree of uncertainty should be indicated at least in qualitative terms. For example, forecasting of droughts 2 or 3 years in advance is not yet possible, although the statistical probability that a drought (of a given severity) will occur sometime in the next hundred years can be estimated with some confidence. The decision-maker should be aware of the degree of uncertainty, which surrounds the predicted state of the environment, and have some understanding of the methods by which this uncertainty is calculated.

Predicting the Future State in the Presence of Action

For each of the proposed actions, and for admissible combinations of these actions, there will be an expected state of the environment, which is to be compared with the expected state in the absence of action. Consequently, predictions similar to those outlined above must be derived for each of the proposed alternatives.

Table presents the main areas of concern that may affect human beings with regard to forecasting the environmental state in the presence of actions:

Table: Areas of Human Concern

Areas of Human Concern (impact categories)	
Economic and Occupational status	Displacement of population; relocation of population in response to employment opportunities; services and distribution patterns; property values.
Social pattern or life style	Resettlement; rural depopulation; change in population density; food; housing; material; agricultural; rural; urban.
Social amenities and relationships	Family life styles; schools; transportation; community feelings; participation vs. alienation; recreation; language.
Psychological features	Involvement; expectations; stress; frustration; Commitment.
Physical amenities (intellectual, cultural, aesthetic and sensual)	National parks; wildlife; art galleries; archaeological monuments; wilderness; clean air and water.
Health	Changes in health; medical services; medical standards.
Personal security	Freedom from molestation; freedom from natural disasters.
Regional and traditional beliefs	Symbols; taboos; values.
Technology	Security; hazards; safety measures; benefits; emission of wastes; congestion; density.
Cultural	Leisure; new values; heritage; traditional and religious rites.
Political	Authority; level and degree of involvement; priorities; structure of decision-making; responsibility and responsiveness; resource allocation; local and minority interests; defence needs.
Legal	Restructuring of administrative management; changes in taxes; public policy.
Aesthetic	Visual physical changes; moral conduct; sentimental values.
Statutory laws and acts	Air and water quality standards; safety standards; national building acts; noise-abatement by-laws.

Note that the nature of impact listed in the table is likely to vary from place to place and from time to time, and there will be overlaps between classes (e.g., health depends in part on economic and occupational status).

So far, we dealt with EIA, which is an indispensable tool for environmental engineers and managers alike. Now, let us introduce you to a new concept called strategic environmental assessment, which covers policies, plans and programmes at critical stages of development.

Strategic Environmental Assessment (SEA)

One of the most recent trends in EIA is its application at earlier, more strategic stages of development at the level of policies, plans and programmes, and is known as strategic environmental assessment (SEA). SEA is defined as the formalised, systematic and comprehensive process of evaluating the environmental impacts of a policy, plan or programme (PPP) and its alternatives, including the preparation of a written report on the findings of that evaluation, and using the findings in publicly accountable decision-making (Therivel, et al., 1992). In other words, the EIA of policies, plans and programmes, keeping in mind that the process of evaluating environmental impacts at a strategic level, is not necessarily the same as that at a project level. In theory, PPPs are tiered – a policy provides a framework for the establishment of plans, plans provide frameworks for programmes and programmes lead to projects. The EIAs for these different PPP tiers can themselves be tiered as shown in Figure, and so the issues at higher tiers need not be reconsidered as the lower tiers:

Tiers in SEAs:

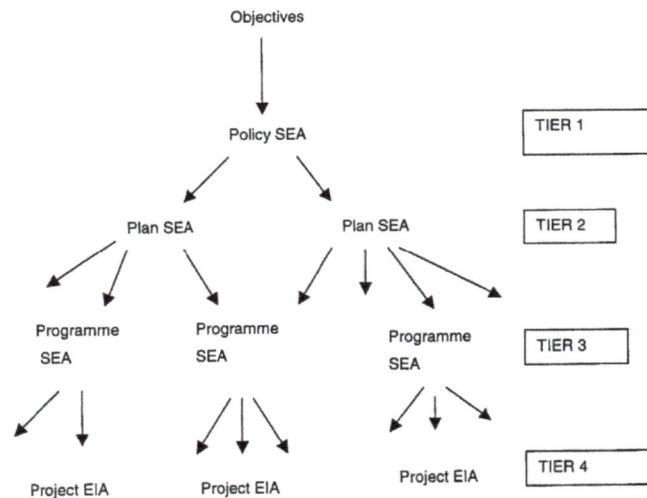

A hierarchy exists between policies, plans and programs with policies are at the top level of conceptualization and generality; plans are one level down from policies, and programs. Programs make plans more specific by including a time schedule for specific activities. Implementation of a program involves carrying out specific projects, which can be subjected to traditional EIA.

Strategic environmental assessment (SEA) is a systematic decision support process, aiming to ensure that environmental and possibly other sustainability aspects are considered effectively in policy, plan and programme making. In this context, following Fischer (2007) SEA may be seen as:

- a structured, rigorous, participative, open and transparent environmental impact assessment (EIA) based process, applied particularly to plans and programmes, prepared by public planning authorities and at times private bodies,

- a participative, open and transparent, possibly non-EIA-based process, applied in a more flexible manner to policies, prepared by public planning authorities and at times private bodies, or

- a flexible non-EIA based process, applied to legislative proposals and other policies, plans and programmes in political/cabinet decision-making.

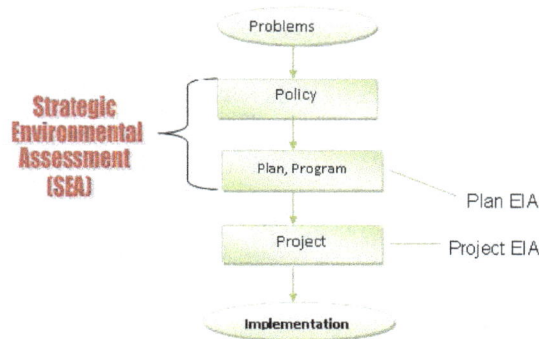

Policy Plan Program (PPP) and S EA

Effective SEA works within a structured and tiered decision framework,aiming to support more effective and efficient decision-making for sustainable development and improved governance by providing for a substantive focus regarding questions, issues and alternatives to be considered in policy, plan and programme (PPP) making.

SEA is an evidence-based instrument, aiming to add scientific rigour to PPP making, by using suitable assessment methods and techniques. Ahmed and Sanchez Triana (2008) developed an approach to the design and implementation of public policies that follows a continuous process rather than as a discrete intervention.

History

The European Union Directive on Environmental Impact Assessments (85/337/EEC, known as the *EIA Directive*) only applied to certain projects. This was seen as deficient as it only dealt with specific effects at the local level whereas many environmentally damaging decisions had already been made at a more strategic level (for example the fact that new infrastructure may generate an increased demand for travel).

The concept of strategic assessments originated from regional development / land use planning in the developed world. In 1981 the *U.S. Housing and Urban Development Department* published the *Area-wide Impact Assessment Guidebook*. In Europe the *Convention on Environmental Impact Assessment in a Transboundary Context* the so-called *Espoo Convention* laid the foundations for the introduction of SEA in 1991. In 2003, the Espoo Convention was supplemented by a Protocol on Strategic Environmental Assessment.

The European SEA Directive 2001/42/EC required that all member states of the European Union should have ratified the Directive into their own country's law by 21 July 2004.

Countries of the EU started implementing the land use aspects of SEA first, some took longer to adopt the directive than others, but the implementation of the directive can now be seen as completed. Many EU nations have a longer history of strong Environmental Appraisal including Denmark, the Netherlands, Finland and Sweden. The newer member states to the EU have hurried in implementing the directive.

Relationship with Environmental Impact Assessment

For the most part, an SEA is conducted before a corresponding EIA is undertaken. This means that information on the environmental impact of a plan can cascade down through the tiers of decision making and can be used in an EIA at a later stage. This should reduce the amount of work that needs to be undertaken. A handover procedure is foreseen.

Aims and Structure of SEA

The SEA Directive only applies to plans and programmes, not policies, although policies within plans are likely to be assessed and SEA can be applied to policies if needed and in the UK certainly, very often is.

The structure of SEA (under the Directive) is based on the following phases:

- "Screening", investigation of whether the plan or programme falls under the SEA legislation,

- "Scoping", defining the boundaries of investigation, assessment and assumptions required,

- "Documentation of the state of the environment", effectively a *baseline* on which to base judgments,

- "Determination of the likely (non-marginal) environmental impacts", usually in terms of Direction of Change rather than firm figures,

- Informing and consulting the public,

- Influencing "Decision taking" based on the assessment and,

- Monitoring of the effects of plans and programmes after their implementation.

The EU directive also includes other impacts besides the environmental, such as material assets and archaeological sites. In most western European states this has been broadened further to include economic and social aspects of sustainability.

SEA should ensure that plans and programmes take into consideration the environmental effects they cause. If those environmental effects are part of the overall decision taking it is called *Strategic Impact Assessment*.

SEA in the European Union

SEA is a legally enforced assessment procedure required by Directive 2001/42/EC (known as the SEA Directive). The SEA Directive aims at introducing systematic assessment of the environmental effects of strategic land use related plans and programs. It typically applies to regional and local, development, waste and transport plans, within the European Union. Some plans, such as finance and budget plans or civil defence plans are exempt from the SEA Directive, it also only applies to plans that are required by law, which interestingly excludes national government's plans and programs, as their plans are 'voluntary', whereas local and regional governments are usually required to prepare theirs.

United Kingdom

SEA within the UK is complicated by different Regulations, guidance and practice between England, Scotland, Wales and Northern Ireland. In particular the SEA Legislation in Scotland (and in Northern Ireland, which specifically refers to the Regional Development Strategy) contains an expectation that SEA will apply to strategies as well as plans and programmes. In the UK, SEA is inseparable from the term 'sustainability', and an SEA is expected to be carried out as part of a wider Sustainability Appraisal (SA), which was already a requirement for many types of plan before the SEA directive and includes social, and economic factors in addition to environmental. Essentially an SA is intended to better inform decision makers on the sustainability aspects of the plan and ensure the full impact of the plan on sustainability is understood.

The United Kingdom in its strategy for sustainable development, *A Better Quality of Life* (May 1999), explained sustainable development in terms of four objectives. These are:

- social progress which recognises the needs of everyone

- effective protection of the environment

- prudent use of natural resources

- maintenance of high and stable levels of economic growth and employment.

These headline objectives are usually used and applied to local situations in order to assess the impact of the plan or program.

SEA Internationally

The Pan-European Region

The Protocol on Strategic Environmental Assessment was negotiated by the member States of the UNECE (in this instance Europe, Caucasus and Central Asia). It required ratification by 16 States to come into force, which it did in July 2010. It is now open to all UN Member States. Besides its potentially broader geographical application (global), the Protocol differs from the corresponding European Union Directive in its non-mandatory application to policies and legislation - not just plans and programmes. The Protocol also places a strong emphasis on the consideration of health, and there are other more subtle differences between the two instruments.

New Zealand

SEA in New Zealand is part of an integrated planning and assessment process and unlike the US is not used in the manner of Environmental impact assessment. The Resource Management Act 1991 has, as a principal objective, the aim of sustainable management. SEA is increasingly being considered for transportation projects.

The OECD DAC - SEA in Development Co-operation

Development assistance is increasingly being provided through strategic-level interventions, aimed to make aid more effective. SEA meets the need to ensure environmental considerations are taken into account in this new aid context. Applying SEA to development co-operation provides the environmental evidence to support more informed decision making, and to identify new opportunities by encouraging a systematic and thorough examination of development options.

The OECD Development Assistance Committee (DAC) Task Team on SEA has developed guidance on how to apply SEA to development co-operation. The document explains the benefits of using SEA in development co-operation and sets out key steps for its application, based on recent experiences.

References

- Finnveden, G.; Moberg, A. (2005). "Environmental systems analysis tools – An overview". Journal of Cleaner Production 13 (12): 1165–1173. doi:10.1016/j.jclepro.2004.06.004

- Baumann, H.; Tillman, A.-M. (2004). The hitch hiker's guide to LCA: An orientation in life cycle assessment methodology and application. Lund, Sweden: Studentlitteratur. ISBN 978-91-44-02364-9

- The Environment Protection and Biodiversity Conservation Act, Australia: The Department of the Environment, Water, Heritage and the Arts, retrieved 9 September 2010

- Udo de Haes, H.; Heijungs, R.; Huppes, G.; Van Der Voet, E.; Hettelingh, J.-P. (2000). "Full mode and attribution mode in environmental analysis". Journal of Industrial Ecology 4 (1): 45–56. doi:10.1162/108819800569285

- Host:Allan McFee (9 May 1977). "The Berger Report is released". As It Happens. transcript. Toronto. CBC Radio. CBC Radio 1. Retrieved 9 December 2011

- Finnveden, G.; Moberg, A. (2005). "Environmental systems analysis tools – An overview". Journal of Cleaner Production 13 (12): 1165–1173. doi:10.1016/j.jclepro.2004.06.004

- Eccleston, Charles H. (2011). Environmental Impact Assessment: A Guide to Best Professional Practices. Chapter 5. ISBN 978-1439828731

- "LCQ1: Judicial review case regarding the Environmental Impact Assessment reports of the Hong Kong-Zhuhai-Macao Bridge". Press Releases. www.info.gov.hk. October 26, 2011. Retrieved October 27, 2011

- Udo de Haes, H.; Heijungs, R.; Huppes, G.; Van Der Voet, E.; Hettelingh, J.-P. (2000). "Full mode and attribution mode in environmental analysis". Journal of Industrial Ecology 4 (1): 45–56. doi:10.1162/108819800569285

- Eccleston, Charles H. and J. Peyton Doub (2010). Preparing NEPA Environmental Assessments: A User's Guide to Best Professional Practices. ISBN 978-1439808825

- Central Environmental Authority. "Environmental Impact Assessment (EIA) Procedure in Sri Lanka". Central Environmental Authority. Central Environmental Authority. Retrieved 1 November 2016

- Finnveden, G.; Moberg, A. (2005). "Environmental systems analysis tools – An overview". Journal of Cleaner Production 13 (12): 1165–1173. doi:10.1016/j.jclepro.2004.06.004

- Eccleston, Charles; Doub, J. Peyton (2012). Preparing NEPA Environmental Assessments: A User's Guide to Best Professional Practices. CRC Press. ISBN 9781439808825

- "Sive,D. & Chertok,M., "Little NEPAs" and Environmental Impact Assessment Procedures" (PDF). Retrieved 2013-01-03

- Udo de Haes, H.; Heijungs, R.; Huppes, G.; Van Der Voet, E.; Hettelingh, J.-P. (2000). "Full mode and attribution mode in environmental analysis". Journal of Industrial Ecology 4 (1): 45–56. doi:10.1162/108819800569285

- Eccleston, Charles H. (2008). NEPA and Environmental Planning: Tools, Techniques, and Approaches for Practitioners, pp 148-149. CRC Press. ISBN 9780849375590

- Finnveden, G.; Moberg, A. (2005). "Environmental systems analysis tools – An overview". Journal of Cleaner Production 13 (12): 1165–1173. doi:10.1016/j.jclepro.2004.06.004

- "Convention on Environmental Impact Assessment in a Transboundary Context (Espoo, 1991)". Unece.org. Retrieved 2013-01-03

- Udo de Haes, H.; Heijungs, R.; Huppes, G.; Van Der Voet, E.; Hettelingh, J.-P. (2000). "Full mode and attribution mode in environmental analysis". Journal of Industrial Ecology 4 (1): 45–56. doi:10.1162/108819800569285

- Eccleston, Charles H. (2014). The EIS Book: Managing and Preparing Environmental Impact Statements. Chapter 6. CRC Press. ISBN 978-1466583630

Environmental Management: Policies and Laws

Environmental policies are formed by organizations or governing bodies to oversee work that concerns the environment. Pesticide and insecticide bioaccumulation, overgrazing, food scarcity and climate change are some of the important issues that are taken up in this field. The major policies and laws of environmental management are discussed in this chapter.

Environmental Policy

Environmental policy refers to the commitment of an organization to the laws, regulations, and other policy mechanisms concerning environmental issues. These issues generally include air and water pollution, waste management, ecosystem management, maintenance of biodiversity, the protection of natural resources, wildlife and endangered species. Policies concerning energy or regulation of toxic substances including pesticides and many types of industrial waste are part of the topic of environmental policy. This policy can be deliberately taken to direct and oversee human activities and thereby prevent harmful effects on the biophysical environment and natural resources, as well as to make sure that changes in the environment do not have harmful effects on humans.

Definition

It is useful to consider that environmental policy comprises two major terms: environment and policy. Environment refers to the physical ecosystems, but can also take into consideration the social dimension (quality of life, health) and an economic dimension (resource management, biodiversity). Policy can be defined as a "course of action or principle adopted or proposed by a government, party, business or individual". Thus, environmental policy focuses on problems arising from human impact on the environment, which retroacts onto human society by having a (negative) impact on human values such as good health or the 'clean and green' environment.

Environmental issues generally addressed by environmental policy include (but are not limited to) air and water pollution, waste management, ecosystem management, biodiversity protection, the protection of natural resources, wildlife and endangered species, and the preservation of these natural resources for future generations. Relatively recently, environmental policy has also attended to the communication of environmental issues.

Rationale

The rationale for governmental involvement in the environment is market failure in the form of forces beyond the control of one person, including the free rider problem and the tragedy of the commons. An example of an externality is when a factory produces waste pollution which may be dumped into a river, ultimately contaminating water. The cost of such action is paid by society-at-large, when they must clean the water before drinking it and is external to the costs of the factory. The free rider problem is when the private marginal cost of taking action to protect the environment is greater than the private marginal benefit, but the social marginal cost is less than the social marginal benefit. The tragedy of the commons is the problem that, because no one person owns the commons, each individual has an incentive to utilize common resources as much as possible. Without governmental involvement, the commons is overused. Examples of tragedies of the commons are overfishing and overgrazing.

Instruments, Problems and Issues

Environmental policy instruments are tools used by governments to implement their environmental policies. Governments may use a number of different types of instruments. For example, economic incentives and market-based instruments such as taxes and tax exemptions, tradable permits, and fees can be very effective to encourage compliance with environmental policy. Corporate companies who engage in efficient environmental management and are transparent about their environmental data and reporting benefit from improved business performance.

Bilateral agreements between the government and private firms and commitments made by firms independent of government requirement are examples of voluntary environmental measures. Another instrument is the implementation of greener public purchasing programs.

Several instruments are sometimes combined in a policy mix to address a certain environmental problem. Since environmental issues have many aspects, several policy instruments may be needed to adequately address each one. Furthermore, a combination of different policies may give firms greater flexibility in policy compliance and reduce uncertainty as to the cost of such compliance.

Government policies must be carefully formulated so that the individual measures do not undermine one another, or create a rigid and cost-ineffective framework. Overlapping policies result in unnecessary administrative costs, increasing the cost of implementation. To help governments realize their policy goals, the OECD Environment Directorate collects data on the efficiency and consequences of environmental policies implemented by the national governments. The website provides database detailing countries' experiences with their environmental policies. The United Nations Economic Commission for Europe, through UNECE Environmental Performance Reviews, evaluates progress made by its member countries in improving their environmental policies.

The current reliance on a market-based framework is controversial, however, and many environmentalists contend that a more radical, overarching approach is needed than a set of specific initiatives, to deal with the climate change. For example, energy efficiency measures may actually increase energy consumption in the absence of a cap on fossil fuel use, as people might drive more fuel-efficient cars. Thus, Aubrey Meyer calls for a 'framework-based market' of contraction and convergence. The Cap and Share and the Sky Trust are proposals based on the idea.

Environmental impact assessments (EIA) are conducted to compare impacts of various policy alternatives. Moreover, it is assumed that policymakers make rational decisions based on the merits of the project. Eccleston and March argue that although policymakers normally have access to reasonably accurate information, political and economic factors often lead to environmentally destructive decisions in the long run.

The decision-making theory casts doubt on this premise. Irrational decisions are reached based on unconscious biases, illogical assumptions, and the desire to avoid ambiguity and uncertainty.

Eccleston identifies and describes five of the most critical environmental policy issues facing humanity: water scarcity, food scarcity, climate change, the peak oil, and the population paradox.

Research and Innovation Policy

Synergic to the environmental policy is the environmental research and innovation policy. An example is the European environmental research and innovation policy, which aims at defining and implementing a transformative agenda to greening the economy and the society as a whole so to achieve a truly sustainable development. Europe is particularly active in this field, via a set of strategies, actions and programmes to promote more and better research and innovation for building a resource-efficient, climate resilient society and thriving economy in sync with its natural environment. Research and innovation in Europe are financially supported by the programme Horizon 2020, which is also open to participation worldwide.

History

The 1960s marked the beginning of modern environmental policy making. Although mainstream America remained oblivious to environmental concerns, the stage had been set for change by the publication of Rachel Carson's New York Times bestseller Silent Spring in 1962. Earth Day founder Gaylord Nelson, then a U.S. Senator from Wisconsin, after witnessing the ravages of the 1969 massive oil spill in Santa Barbara, California. Administrator Ruckelshaus was confirmed by the Senate on December 2, 1970, which is the traditional date used as the birth of the agency. Five months earlier, in July 1970, President Nixon had signed Reorganization Plan No. 3 calling for the establishment of EPA in July 1970. At the time, Environmental Policy was a bipartisan issue and the efforts of the United States of America helped spark countries around the world to create environmental policies. During this period, legislation was passed to regulate pollutants that go into the air, water tables, and solid waste disposal. President Nixon signed the Clean Air Act in 1970 which set the USA as one of the world leaders in environmental conservation.

In the European Union, the very first Environmental Action Programmed was adopted by national government representatives in July 1973 during the first meeting of the Council of Environmental Ministers. Since then an increasingly dense network of legislation has developed, which now extends to all areas of environmental protection including air pollution control, water protection and waste policy but also nature conservation and the control of chemicals, biotechnology and other industrial risks. EU environmental policy has thus become a core area of European politics.

Overall organizations are becoming more aware of their environmental risks and performance requirements. In line with the ISO 14001 standard they are developing environmental policies suitable for their organization. This statement outlines environmental performance of the organization as well as its environmental objectives. Written by top management of the organization they document a commitment to continuous improvement and complying with legal and other requirements, such as the environmental policy objectives set by their governments.

Environmental Policy Integration

The concept of environmental policy integration (EPI) refers to the process of integrating environmental objectives into non-environmental policy areas, such as energy, agriculture and transport, rather than leaving them to be pursued solely through purely environmental policy practices. This is oftentimes particularly challenging because of the need to reconcile global objectives and international rules with domestic needs and laws. EPI is widely recognised as one of the key elements of sustainable development. More recently, the notion of 'climate policy integration', also denoted as 'mainstreaming', has been applied to indicate the integration of climate considerations (both mitigation and adaptation) into the normal (often economically focused) activity of government.

Environmental Policy Studies

Given the growing need for trained environmental practitioners, graduate schools throughout the world offer specialized professional degrees in environmental policy studies. While there is not a standard curriculum, students typically take classes in policy analysis, environmental science, environmental law and politics, ecology, energy, and natural resource management. Graduates of these programs are employed by governments, international organizations, private sector, think tanks, universities, and so on.

Due to the lack of standard nomenclature, institutions use varying designations to refer to academic degrees they award. However, the degrees typically fall in one of four broad categories: master of arts, master of science, master of public administration, and PhD in environmental policy. Sometimes, more specific names are used to reflect the focus of the academic program. For example, the Middlebury Institute of International Studies at Monterey awards master of arts in international environmental policy (MAIEP) to emphasize the international orientation of the curriculum.

Understanding Environmental Policies

Environmental policies may be either enacted as laws by governing bodies or created and enforced by government agencies. They may originate from local, national or foreign governments, and address an array of issues including (but not limited to) air or water quality, fossil fuel extraction, energy conservation, habitat protection or restoration, pesticide use, storage/disposal of hazardous materials, recycling and trafficking in endangered species.

An environmental policy being interdisciplinary in nature draws together technology, economics, and natural and social sciences. In order to develop sustainable policies, therefore, it is necessary to have sound knowledge of the actual and potential environmental impacts of certain activities and some knowledge of the technical characteristics, economic costs, social acceptability and possible side effects of alternative policy options.

The quality of the environment has both a direct and an indirect effect on the standard of living. This does not mean that environmental degradation is simply a by-product of economic activities, it is also the consequence of the priorities set by States in their economic policies. These policies generally aim at stimulating production and, as a consequence, tend to ignore their implications for the environment. Past experience, however, shows that economic policies may actually have more impact on the quality of the environment than those policies explicitly designed to protect the environment. We will discuss this, next.

Economics and Environmental Policies

One sign of a sustainable economy is when the costs of environment and health caused by economic growth have been added to consumer prices and when economic policy instruments support sustainable development. Environmental policies should supplement economic instruments.

Environmental policies involve certain measures aimed at achieving a sound environment. They are usually developed in the context of public policy, based on economic theory, which focuses more on the level of costs and benefits associated with the implementation of environmental policies than on the quality of the environment. When governments propose and subsequently implement strict standards, sectors that pollute the environment will have to take measures, and this cannot be achieved without incurring extra costs. Polluting industries are, therefore, often keen to highlight the likely costs they have to incur due to the proposed environmental measures.

In other words, the immediate benefits resulting from environmental policies are extremely difficult to assess. As a consequence, the costs of environmental measures are often paid more attention than the benefits resulting from the implementation of the policy.

The definition of the property rights of natural resources plays a vital role in the distributional effects of environmental policies. The implementation of strict standards and regulations will effect a change in the definition of property rights. For example, industries polluting the rivers will be confronted with regulations that prevent them from, or reduce their opportunities for, using the rivers. However, throughout the process of formulating the regulations, polluting industries will try to influence and stifle the policies.

Let us now consider below a few examples of sectoral economic policies that influence the environmental policies directly or indirectly:

- Agricultural sector: Virtually the entire food cycle attracts huge direct or indirect subsidies, at a cost to taxpayers and consumers. These subsidies, more often than not, send farmers far more powerful signals than do the small grants, usually provided for soil and water conservation. They encourage farmers to occupy marginal land and to clear forests and woodlands, make excessive use of pesticides and fertilisers, and use underground and surface waters in irrigation indiscriminately.

- Forestry sector: The pressures on forests throughout the world vary greatly in both developed and developing countries, which are reinforced by government policies. The logging and forestry industry attracts a variety of direct and indirect subsidies. The perverse incentives that encourage the over harvesting of temperate as well as tropical forests also mark world-trade in forest products.

- Transport sector: This sector, especially motor vehicles, also benefits from economic policies that are ecologically perverse. Fuel taxes in many jurisdictions, for example, still fail to distinguish between the environmental effects of different types of fuel (e.g., petrol or diesel, leaded or unleaded). The tax and tariff structure, and direct and indirect subsidies, encourage heavier and more energy-intensive vehicles and road freight, as opposed to rail transport in many countries. In addition, in some countries, private vehicle expenses can be deducted from taxable income

- Energy sector: The major obstacle to energy efficiency is the existing framework of incentives for energy exploration, development and consumption. These incentives underwrite coal, oil and gas, ignore the costs of air, land and water pollution and seem to favour inefficiency and waste. While industrialised countries have been spending billions to distort the market and consumer prices in ways that actively promote acid rain and global warming, they have been spending only a few million on measures to promote energy efficiency. As long as pollution problems are mainly national, there is a need for a strong national authority. However, environmental problems are becoming increasingly international or global. This complicates the environmental policies considerably. On the one hand, international co-operation in the fighting of environmental problems is absolutely necessary. On the other hand, different countries have different economic interests. Furthermore, polluting sectors are not evenly distributed among countries.

Economic based environment policies have been designed to facilitate economic growth and allow business while ensuring the sustainability of the environment and achieve economic efficiency.

Industries and Environmental Policies

Industries are a measure of a country's economic growth. Consequently, countries have a tendency to protect their polluting industries, in particular when they are relatively important economically. However, the growing interest in environmental management has fuelled certain industries to adopt policies that are economically feasible and which helps curb environmental degradation.

Various factors drive the development of a managed approach to environmental performance. These include the following:

- The need to meet increasingly stricter environmental regulations.

- Stakeholder pressure (e.g., pressure from shareholders, insurers and investors).

- Supply-chain pressure from customers.

- Historically poor relations with regulatory bodies and local communities

Many industries have established environmental management systems (EMS) to tackle activities, which either pose a serious threat to ecosystems in the event of accidents or involve significant expenditure because of the costs associated with raw material use and/or waste disposal. An

EMS is "the part of an overall management system that includes organisational structure, planning activities, responsibilities, practices, procedures, processes and resources for developing, implementing, achieving reviewing and maintaining the environmental policy". A EMS aims to help organisations achieve sound environmental performance by identifying key activities which impact, already or potentially, on the environment and by putting in place management controls to ensure that the organisation continues to meet its legal and policy requirements to deal with these impacts.

Traditionally, most of the work on EMS has been done by those industrial sectors with the greatest potential to affect the environment, e.g., the chemical, waste management and oil refining sectors. In addition to this basic need to control pollution, some companies have identified EMS as an effective way of improving productivity through waste and resource minimisation initiatives and a mechanism for increasing sales as customers turn on to environmentally safer products.

Arguably, implementing effective EMS can be a useful deregulatory tool for business. For example, by reducing the use of certain hazardous substances, some organisations in India have achieved major improvements in their air emissions and, as a result, they came out of the strict controls imposed on them by the Environmental Protection Act (EPA) 1990. In addition, they have gained a better working environment for their employees and eliminated a difficult raw material storage hazard.

Close on the heel of industrial policy is the agricultural policy that can be adopted to prevent the deleterious effect of agricultural activities on the environment.

Agriculture and Environmental Policies

Agriculture has a major impact on the environment, especially on land, water and biodiversity. Over the last 10 - 15 years, the environmental performance of agriculture has been mixed. For example, nitrogen and pesticide loading in water remain relatively high and risks of soil erosion and water resource depletion persist in many regions and countries. In recent years, however, there have been improvements in wildlife habitats, landscapes and sinks for greenhouse gases provided by agriculture, but the most significant progress has occurred where environmental pressures have been greatest.

The main environmental impacts of agriculture may be characterised through the beneficial or harmful contribution of agricultural activities to:

- soil quality (e.g., erosion, nutrient supply, moisture balance, salinity, etc.);

- land quality (e.g., ecological management of agricultural land);

- water quality (e.g., nutrient, pesticide and sediment run-off and leaching, salinity);

- water quantity (e.g., irrigation consumption, use efficiency, water retention capacity, flood prevention, etc.);

- air quality (e.g., emissions of dust, odours, ammonia and greenhouse gas, absorption of carbon dioxide, etc.);

- bio-diversity (e.g., farm and indigenous animal and plant diversity);

- wildlife and semi-natural habitats (e.g., diversity of animal and plant habitats associated with farming);

- rural landscape (e.g., environmental features of areas shaped by farming, including those associated with historic buildings and landmarks, etc.).

Agricultural policies in India provide substantial farm support, often linked to commodity production affecting resource use, farming practices and environmental performance. Reconciling food production and environmental goals, however, is a challenge. But, reconciling them implies that the rights and responsibilities of farmers regarding farm practices need to be clearly defined and applied, and thus the situations under which they are entitled to remuneration or obliged to pay (polluter-pays-principle or PPP). Defining who pays and who is paid for the desired level of environmental performance has important implications for the distribution of income and wealth.

Role of Agro-environmental Policies

When private and public mechanisms designed to facilitate the improvement and diffusion of appropriate farming practices and market forces are not enough to ensure the supply to meet the demand of environmental services, specific agricultureenvironmental measures at the farm level may be needed. Such measures may be necessary to reduce the environmental harm, or enhance the environmental benefits of farming activities. When designing and implementing such policy measures, a number of general policy principles should be taken into account in the choice of the type of policy incentive or disincentive – payment or tax.

The PPP applies to reducing environmental harm for which farmers, as any other polluter, should be accountable. However, the PPP guiding principles recognise the possibility of different property rights and reference levels among countries, with the possibility of offering transitional financial incentives to encourage farmers to adopt appropriate production practices for improving their environmental performance through reducing environmental harm. It includes the case of transitional financial assistance provided to stimulate the development of new pollution control technologies and abatement equipment to achieve a better environmental performance through improved production practices.

General Policy Principles

- When markets do not exist to allocate costs and benefits of agriculture-environmental impacts and outputs, policy action may be needed to account for the costs of not respecting environmental targets and to ensure the provision of environmental benefits. When designing and implementing policy measures, the environmental problem needs to be clearly defined and the following principles for policy design need to be kept in mind:

- The necessary condition for a welfare gain from implementing an agriculture-environmental policy measure is that the resulting environmental benefits exceed the costs associated with the policy. These costs include those due to a reduction in outputs, associated with more environmentally friendly technologies and practices and the transaction (administrative) costs of policy implementation and enforcement.

- When farmers and other economic agents provide a specific environmental service, the level of benefit should be clearly specified and efforts made to ensure that the most efficient operator is the provider.

- When a specific environmental outcome is jointly the result of agricultural output, a wide range of policy options and approaches may achieve its provision by an individual farmer that either provides positive incentives (through, for example, a payment) or negative incentives (e.g., a tax). If incentives were set correctly, it would be in the individual farmer's interest to achieve the outcome and receive the incentive payment, or achieve the outcome and avoid paying the tax.

- The effectiveness of either a tax or a payment depends not only on whether it correctly confronts the farmer with the opportunity costs of not respecting environmental requirements, but also on the degree to which the associated obligations can be enforced and tailored to local environmental circumstances and demands. The more the payment or tax is tailored to specific circumstances, the larger the need for monitoring, the lower the probability of individual control, and the higher the transaction costs.

In addition to what we have discussed, an environmental policy that conserves the ecosystem is a sagacious attempt to manage the environment.

Ecosystem and Environmental Policies

The objective of ecosystem conservation and management as stipulated in the Forest Policy (MNRT, 1998a) is to ensure ecosystem stability through conservation of forest biodiversity, water catchments and soil fertility. Forest biodiversity is faced with problems of encroachment, shifting cultivation, wild fires, lack of systematic forest management and inadequate infrastructure and staff to prevent excessive resource use.

Watershed management and soil conservation face similar problems as biodiversity conservation due to increase in population pressure and inefficient forest management and protection in watershed areas. Poor management and protection have resulted in deterioration of watershed areas causing water shortages. Planting of inappropriate species in watershed areas, illegal logging and inappropriate logging methods have further reduced the quantity and quality of water and are the cause of peak floods, droughts and sedimentation in rivers. Erosion due to cultivation on the riverbanks outside forest reserves is also a major concern in watershed management.

Encroachment, wildfires, illegal logging and poaching are also the major factors contributing to the deterioration of wildlife populations in natural forests. Inadequacy of baseline data on types of wildlife species, their habitats and the lack of incorporating wildlife management in forest management plans, are the major reasons hampering wildlife conservation in forest reserves. Also, coordination among the government institutions involved in wildlife and forest management is very poor.

In order to minimise the damage to the ecosystem due to human activities, the Forest Policy stipulates environmental impact assessment (EIA) before implementing developmental projects, which convert forest land to other land uses such as timber industries, mining, road construction, agriculture, dams, settlements, shrimp farming and tourism. The scope and guidelines of a EIA is to be prepared in collaboration with other sectors and stakeholders.

Specifically, in order to achieve ecosystem conservation and management in the areas of biodiversity, watershed management and soil conservation, and wildlife, we need to consider:

- New forest reserves for biodiversity conservation in areas of high biodiversity value. Forest reserves with protection objectives of national strategic importance may be declared as nature reserves.

- Biodiversity conservation and management is to be included in the management plans for all protection of forests. Involvement of local communities and other stakeholders in conservation and management is to be encouraged through joint management agreements.

- Biodiversity research and information dissemination should be strengthened in order to improve biodiversity conservation and management.

- Biodiversity conservation must be incorporated in the management regimes of natural production forests and plantations. Management plans must incorporate biodiversity conservation and management guidelines. This minimises the replacement of natural forests by exotic plantations.

- Watershed management and soil conservation should be included in the management plans for all protection and production forests. Involvement of local communities and other stakeholders in watershed management and soil conservation will be encouraged through joint management agreements.

- Research and information dissemination must be strengthened in order to improve watershed management and soil conservation.

- EIA is required for the investments, which convert forestland to other land use or may cause potential damage to the forest environment.

Having looked at some of the sectoral policies that influence environment policies, let us now discuss the instruments used in implementing these policies.

Environmental Policy Instruments (EPI)

Environmental policy instruments (EPI) usually refer to official actions taken to curb and remove the negative environmental impacts caused by society. The methods, laws, administration and decisions relating to these actions are collectively termed environmental policy. A EPI is divided into economic, information and legal measures. Indicators of sustainable development frequently measure the status of development or pressures directed at it. Indicators of environmental policy instruments reflect society's reaction, and the steps taken to make sustainable development possible.

A number of different typologies are used to classify EPI. One of the most commonly used typologies divides a EPI into the following three categories:

(i) Regulatory instruments that mandate specific behaviour.

(ii) Market-based instruments that act as incentives for particular activities.

(iii) Information-based instruments that seek to change behaviour through the provision of information.

We will discuss each of these, next. But, let us first note that governments may establish formal cleaner production strategies or programmes to act as a framework for the coordinated implementation of subsequent and more specific policy instruments. Cleaner production strategies may take one or a combination of the following shapes:

Product bans: The imposition of a ban or defined phase-out schedule for a particular product or substance is an authoritarian means of promoting cleaner production. This may be implemented through application of the product choice or substitution principle.

Extended producer responsibility (EPR): EPR aims at making environmental improvements throughout the life cycle of a product by making the manufacturer responsible for various aspects of the product's life cycle. In particular, this could include the take-back recycling and final disposal of the product.

Cleaner production audits: As part of their permitting requirements, it is mandatory for production industries to carry out cleaner production audits of their plants and to implement findings as long as they do not harm the environment.

Mandatory EMS and reporting: In terms of integrated permit conditions, it is mandatory for production industries to implement a structured environmental management system and make public information on their environmental performance.

 Financial and technical incentives: Governments may stimulate cleaner production measures by providing grants, loans and favourable tax regimes, and/or by supplying targeted technical assistance to relevant industrial enterprises.

Regulatory Instruments

Since the inception of environmental policy, the predominant strategy for pollution control has generally been through the use of regulatory instruments. Usually, a public authority sets standards, and then inspects, monitors and enforces compliance to these standards, punishing transgressions with formal legal sanction. These regulations may, for example, specify an environmental goal such as the reduction of carbon dioxide emissions by a specified date. They may also mandate the use of a particular technology or process. Such an approach gives the regulator the maximum authority to control where and how resources will be allocated to achieve environmental objectives. Also, this provides the regulator with a reasonable degree of predictability as to how much the pollution levels will be reduced.

Market-based Environmental Policy Instruments

In environmental law and policy, market-based instruments (MBIs) are policy instruments that use markets, price, and other economic variables to provide incentives for polluters to reduce or eliminate negative environmental externalities. MBIs seek to address the market failure of externalities (such as pollution) by incorporating the external cost of production or consumption activities through taxes or charges on processes or products, or by creating property rights and facilitating the establishment of a proxy market for the use of environmental services. Market-based instruments are also referred to as economic instruments, price-based instruments, new environmental policy instruments (NEPIs) or 'new instruments of environmental policy.

Examples include environmentally related taxes, charges and subsidies, emissions trading and other tradeable permit systems, deposit-refund systems, environmental labeling laws, licenses, and economic property rights. For instance, the European Union Emission Trading Scheme is an example of a market-based instrument to reduce greenhouse gas emissions.

Market-based instruments differ from other policy instruments such as voluntary agreements (actors voluntarily agree to take action) and regulatory instruments (sometimes called "command-and-control"; public authorities mandate the performance to be achieved or the technologies to be used). However, implementing an MBI also commonly requires some form of regulation. Market based instruments can be implemented in a systematic manner, across an economy or region, across economic sectors, or by environmental medium (e.g. water). Individual MBIs are instances of environmental pricing reform.

According to Kete (2002), "policymaking appears to be in transition towards more market-oriented instruments, but it remains an open-ended experiment whether we shall successfully execute a long-term social transition that involves the private sector and the state in new relationships implied by the pollution prevention and economic instruments rhetoric."

History

For example, although the use of new environmental policy instruments only grew significantly in Britain in the 1990, British Prime Minister David Lloyd-George may have introduced the first market-based instrument of environmental policy in the UK when a Fuel tax was levied in 1909 during his ministry.

Transferable Permits

A market-based transferable permit sets a maximum level of pollution (a 'cap'), but is likely to achieve that level at a lower cost than other means, and, importantly, may reduce below that level due to technological innovation.

When using a transferable-permit system, it is very important to accurately measure the initial problem and also how it changes over time. This is because it can be expensive to make adjustments (either in terms of compensation or through undermining the property rights of the permits). Permits' effectiveness can also be affected by things like market liquidity, the quality of the property right, and existing market power. Another important aspect of transferable permits is whether they are auctioned or allocated via grandfathering.

An argument against permits is that formalising emission rights is effectively giving people a license to pollute, which is believed to be socially unacceptable. However, although valuing adverse environmental impacts may be controversial, the acceptable cost of preventing these impacts is implicit in all regulatory decisions.

Taxes

A market-based tax approach determines a maximum cost for control measures. This gives polluters an incentive to reduce pollution at a lower cost than the tax rate. There is no cap; the quantity of pollution reduced depends on the chosen tax rate.

A tax approach is more flexible than permits, as the tax rate can be adjusted until it creates the most effective incentive. Taxes also have lower compliance costs than permits. However, taxes are less effective at achieving reductions in target quantities than permits. Using a tax potentially enables a double dividend, by using the revenue generated by the tax to reduce other distortionary taxes through revenue recycling. There can also be conflict between objectives with a tax: less pollution means less revenue.

Market-based vs Command and Control

An alternate approach to environmental regulation is a command and control approach. This is much more prescriptive than market-based instruments. Command and control regulatory instruments include emissions standards, process/equipment specifications, limits on input/output/discharges, requirements to disclose information, and audits. Command and control approaches have been criticised for restricting technology, as there is no incentive for firms to innovate.

Market-based instruments do not prescribe that firms use specific technologies, or that all firms reduce their emissions by the same amount, which allows firms greater flexibility in their approaches to pollution management. However, command and control approaches may be beneficial as a starting point, when regulators are faced with a significant problem yet have too little information to support a market-based instrument. Command and control approaches can also be preferred when regulators are faced with a thin market, where the limited potential trading pools mean the gains of a market-based instrument would not exceed the costs (a key requirement for a successful market-based approach).

Market-based instruments may also be inappropriate in dealing with emissions with local impacts, as trading would be restricted to within that region. They may also be inappropriate for emissions with global impacts, as international cooperation may be difficult to attain.

For a variety of reasons, environmental advocates initially opposed the use of market-based instruments except under very constrained conditions. However, after the successful use of freely traded credits in the lead phasedown in the U.S. environmental advocates recognized that trading markets has benefits for the environment as well. Thereafter, beginning with the proposal of the acid rain allowance market, environmental advocates have supported the use of trading in a variety of contexts.

Market-based instruments generally seek to address the market failure of environmental externalities either by incorporating the external cost of a firm's polluting activities into the firm's private cost (for example, through taxation), or by creating property rights and facilitating the establishment of a proxy market (for example, by using tradable pollution permits).

Before introducing any new economic instruments, governments should identify and evaluate any economic incentives that may already be in operation, either explicitly or implicitly. These include, for example, the use of subsidies to make local industries more competitive. Many of these policies lead to artificially low prices for resources, such as energy and water, and as a result of which these resources may be overused, creating both pollution and shortages. Government assessments of such policies are, therefore, needed before other economic instruments are applied.

Taxes, fees and charges may be used to promote cleaner production practices by raising the costs of unwanted outputs or by providing incentives to promote more efficient use of natural resources. In some instances, it may be appropriate to use the revenues generated from these instruments to support cleaner production activities and thereby stimulating preventative approaches. A significant constraint against the more widespread adoption of market-based instruments, however, is that it is not always politically feasible to set taxes at a sufficiently high level to achieve desired environmental goals. Governments often face resistance, if taxation related to environment is taken merely as a means of increasing its revenues.

Governments may be able to avoid some obstacles by earmarking the corrected charges or shifting tax sources. In any case, the successful implementation of such instruments requires a system of monitoring, revenue collection and enforcement as well as measures to combat possible corruption.

Financial subsidies, (e.g., low-interest loans, direct grants or preferential tax treatment) can be targeted to specific industries to stimulate technological development. Governments must, however, carefully examine how subsidies work to ensure that they are not misused resulting in environmentally counterproductive behaviour.

Information-based Strategies

In addition to creating an appropriate regulatory and financial framework for cleaner production, government may further stimulate the adoption of cleaner production practices through the use of informational measures. These may be used to provide the right incentive (e.g., through the public disclosure of a firm's environmental performance) as well as to build capacity within industry (e.g., through the publication and dissemination of relevant case studies).

A few examples of information-based strategies are given below:

- Promoting the adoption of targeted, high profile demonstration projects, to demonstrate the techniques and cost-saving opportunities associated with cleaner production.

- Encouraging educational institutions to incorporate preventative environmental management within their curricula, particularly within engineering and business courses.

- Requiring public disclosure of information on environmental performance by, for example, establishing a pollutant release and transfer register, stimulating greater voluntary corporate reporting and requiring the provision of information on specific materials.

- Initiating and/or supporting measures that address consumption such as eco-labelling schemes and environmental product declarations.

- Promoting the adoption of effective training initiatives.

- Issuing high profile awards for enterprises that have effectively implemented cleaner production.

Note that in some instances, a EPI is characterised by more than one of the above categories. Furthermore, it is important to note that different policy instruments are sometimes best used in conjunction with others.

Environmental Policies And Programmes In India

The year 1972 marks a watershed in the history of environmental management in India. Prior to 1972, different government ministries dealt with environmental concerns such as sewage disposal, sanitation and public health, and each pursued these objectives in the absence of a proper coordination system at the governmental or the intergovernmental level. During the twentyfourth UN General Assembly of 1972, Mr. Pitamber Pant, a member of the Planning Commission of India prepared three reports about India. With the help of these reports, the impact of the population explosion on the natural environment and the existing state of environmental problems were examined.

The Planning Commission subsequently set up an expert committee to formulate long-term sectoral (including environment and forest) policies. It also noted that many environmental problems were continuing to cause serious concern, for example, the loss of top soil and vegetative cover, the degradation of forests, continuing pollution by toxic substances, careless industrial and agricultural practices and unplanned urban growth. It acknowledged that environmental degradation was seriously threatening the economic and social progress of the country.

The continuing decline in the quality of the environment spurred the Union, government and a few State governments to adopt stronger environmental policies, to enact fresh legislation and to create, reorganise and expand administrative agencies. Based on the recommendations of the Union Government in April 1990, the Government of India adopted a National Conservation Strategy (NCS) and Policy Statement on Environment and Development.

The preamble to the NCS adopts the policy of sustainable development and declares the government's commitment to reorient policies and action in unison with the environmental perspective. The NCS proceeds to recognise the enormous dimensions of the environmental problems facing India and declares strategies for action in various spheres such as agriculture, forestry, industrial development, mining and tourism.

In February 1992, the Union Government published its policy for the abatement of pollution. This statement declares the objective of the government to integrate environmental considerations into decision-making at all levels. To achieve this goal, the statement adopts some fundamental guiding principles: prevention of pollution at source, adoption of the best available technology, the polluter pays principle and public participation in decision-making.

Thus, the policies and legislation for environmental management in India has evolved over the years to meet the needs of forest conservation for sustainable development and environmental protection for improvement and management of ambient quality of air, water and soil.

Forest Conservation Activities

India is well known for its aranya sanskriti or forest culture. Forest dwelling communities lived in symbiosis with a large number of wild animals, hunting some and venerating others. With the advance of agriculture with the Iron Age, forests began their retreat to the hilly regions of the Himalayas and their lesser ranges, the central Indian plateaus and Western and Eastern ghats. The practices of indigenous forest tribes and people of the Northern and North-Eastern hills still demonstrate the close relationship that formerly existed among human beings, forest and other species.

The gradual decline of forests led to a number of seasonal, ritual and specific restrictions on forest-based activities. They were reinforced by religious and totemic proscriptions against harming certain types of trees and birds, snakes and animals. India's forest tribes still worship a number of trees and natural phenomena. For example, the annual hunt of the Santhals is preceded by an invocation to the deities of the forests and animals. Even the act of cutting the branches of certain trees must be accompanied with a verbal, personally communicated apology to the sacred tree.

The destruction of last reserves of natural forests, however, has its roots in the British colonial policy towards the commercial exploitation of forests. With the introduction of the comprehensive Indian Forest Act in 1878, the colonial state radically redefined proprietary rights, imposing on the forest a system of management and control, the priorities of which sharply conflicted with earlier systems of local use and control. The clear felling of extensive natural forests and introduction of scientific management for their commercial exploitation marked an ecological, economic watershed in Indian forestry history.

The Forest Act of 1878 abolished traditional rights and customary practices governing forest management, and the Indian Forest Act of 1927 gave total control over forest resources to the State. This process continued in independent India, which retained the forest laws of the colonial government. The objective of maximising revenues from the sale of what was once common property resource was now done in the name of national interest and economic growth.

The Forest Policy of 1952, subsequently, emphasised the exploitation of forests primarily for producing valuable timber for industry and other vital purposes of national interest. The policy also strongly emphasised that proximity to forest resources did not ensure village communities rights over the property. The National Commission on Agriculture, 1976, drew up a second policy in favour of commercial viability of forests and production of wood for industrial purposes.

These two policies led to a large reduction in the natural forest cover. This was mainly due to the very low concessional rates charged to industries for raw material. The major negative effects were the non-regeneration of forests due to unsustainable commercial exploitation, denial of peoples' legitimate rights to forest products and imbalance in ecologically sensitive regions leading to land-slides, soil erosion and loss of biodiversity.

NGO movements for Environmental Protection in India

The urgency of industrial development in the urban areas often ignored the needs of the environment, and the feeble voices of the local community were drowned in the publicity of large organisations. As a result, several committed individuals came forth and formed NGO to give voice to the feeble. Let us touch upon some key NGO movements in India.

Silent Valley struggle

In the 1970s, led by Kerala Sastra Sahitya Parishad, students and a galaxy of intellectuals, academics, artists and naturalists conducted a nation-wide movement against a hydroelectric project that would have destroyed part of the Silent Valley, a unique tract of 9,000 hectares of forest in Western Ghats. In those times, none of the government departments/agencies was competent to deal with this kind of public opposition to developmental projects endangering the environment. Ultimately, decisions were taken at the level of the Prime Minister to stop the construction of the hydroelectric dam project. The discovery of nine previously unknown plant species and a new genus by the Botanical Survey of India in the Silent Valley reinforced the activists' determination to protect the forest.

Chipco Movement

In the 1970s, a disastrous flood occurred in the Alaknanda river in Garhwal. Local people were the first to link the disaster with the ecological change that had taken place in the watershed due to felling of forests. As a result, a movement was born in March 1973 in the tiny hamlet of Gopeshwar in the Garhwal region. Here, the villagers were denied access to ash wood for their agricultural implements. When the trees were allotted to a sports good manufacturing company, villagers rushed to protect the same trees from contractor's axe by hugging them. The method of protecting trees by hugging them spread like wildfire throughout the Garhwal region in the next few years. The movement, first interpreted as a fight for people's rights over scarce common property resources, quickly became a true ecological movement, in that it recognised the importance of forests in maintaining the fragile ecosystem. Garhwal villagers asserted that forests were the foundation of both their cultural and material life and, therefore, their birthright. Religious leaders supported the Chipco movement and Gandhian disciples spearheaded its manifestations. There were also folklore reflecting the spiritual and ecological foundation of the Satyagraha.

Chipco succeeded in changing the official strategy of forest exploitation to one responsive to the needs of people. By 1977, it became a full-fledged ecological movement, recognised at the highest level in the policy-making. A moratorium was placed on commercial green felling in the Himalayan region of Uttar Pradesh. Following this, Mr. Sunderlal Bahuguna and other satyagrahis conducted a 4,780 km long march from Garhwal to Kohima to spread the Chipco message. In Garhwal, the Dashauli Gram Swarajyasangh organised the largest voluntary afforestation programme in the country, planting millions of trees to restore the barren hillsides.

Public Interest Litigation for the Protection of Taj Mahal

Mathura Refinery is located in the vicinity of the Taj Mahal in Agra, despite a joint-Parliamentary Committee expressing its serious concerns about the location of the refinery within the Taj Trapezium, a 10,400 sq km area. The Mathura refinery and more than 2,300 small and medium industries around it, including iron foundries, have continuously been emitting tonnes of pollutants such as sulphur dioxide and other poisonous gases, yellowing the Taj Mahal. Environmental Advocate, Mr. M. C. Mehta, filed a public interest litigation in the Supreme Court for the protection of the Taj from serious pollution. A bench of the Supreme Court has passed a number of orders directing the industries to install devices to control pollution or else face closure, the Mathura Refinery to minimize sulphur dioxide emissions and the Union Ministry of Environment and Forests to take suitable measures to save the Taj. Apex Court has given various directions including banning the use of coal and coke and directing the industries to switch over to Compressed Natural Gas (CNG)

So far, we have been discussing the various environmental policies aimed at the protection of the environment. Let us now discuss the legal aspects of environmental management.

Environmental Laws And Legislations

Environmental law refers to rules and regulations governing human conduct likely to affect the environment. It reflects the legislative measures, and the administrative and judicial structures to protect the environment.However, it is difficult to define precisely the boundaries of environmental law in the same way as we define, say, the law of contract explained in detail. Unlike the traditional legal subjects such as contract, which are well developed, environmental law is still in its infancy. Nevertheless, attention is now increasingly focused on the rationalisation and streamlining of existing measures rather than the development of substantial law.

Environmental law aids in:

- regulation of resource use;

- protection of the environment and biodiversity;

- mediation, conflict resolution and conciliation;

- formulation of stable, unambiguous undertakings and agreements.

Legislations have evolved in response to problems, so that there is often a delay between the need and the establishment of satisfactory law. Without effective legislation, resource use, pollution control, conservation and most fields of human activity are likely to fall into chaos and conflict. Law can encourage satisfactory performance, enable authorities to punish those who infringe environmental management legislation, confiscate faulty equipment or close a company. It may also be possible for employees, bystanders and product or service users to sue for damages, if they are harmed. Environmental laws, in essence, are indispensable instruments in curbing environmental degradation. Environmental laws can be categorised as public and private laws depending on the environmental issue.

Private and Public Law

Based on the environmental nuisance cases the legal actions are taken to understand to which category the environmental issue relates to; there is a need to know about private and public law

Private law includes law of contracts, law of torts, law of property and the law of obligations, is the part of legal system that involves relationships between individuals. It is a private BILL enacted into law which is applied to individual or corporation. Labour law, commercial law, corporation's law and competition law come under private law. This law when observed as common law shows relationship between governments and private individuals. Law of contract are governed by private law that affects the relationships between the individuals without the intervention of the state or government.

Public law is a theory of law governing the relationships individuals and the state. The major sub divisions of this law are constitutional law, administrative law and criminal law. For most lawyers and clients this law is how public authorities make decisions. That explains the irrelevant and relevant factors considered by public for which reasons may be important or not. They also analyse whether the decision maker complied with all legal requirements along with EC law regulations to make a decision. This law governs activities of public bodies such as environment agency and regulates the relationship between state and individuals.

Private law can be used in environmental cases when the claimer is against someone causing a nuisance. This nuisance can be causing personal injury, trespassing on the land etc. In these cases it is better to be familiar with private law and environmental law with science background to win the case and to judge which ones are not worth pursuing the case.

Environmental law is practised in the public interest for public benefit in the form of groups or individuals seeking environmental protections. Strength of law lies in its potential elasticity that represents the basis for developing an environmental tort action unconnected to land and capable of protecting wider community interests.

Public interest rarely happens whereas the law that is practised in private interest in the form of groups or individuals who are responsible for polluting or committing environmentally destructive activities who can also avoid violating these environmental laws in the process are usually common cases issues. This clearly shows private law plays negative role in environmental cases. This law balances only individual interests such as challenging uses of land rather than environmental protection.

Principles of International Law

International law governs relations between States and has no direct effect on domestic law or individuals. It is often difficult to force a sovereign State to sign and then honour a treaty or similar agreement. International law, thus, depends a great deal on voluntary agreements by governments and international bodies.

The 1972 UN Conference on the Human Environment (held in Stockholm) is rightly regarded as the starting point of the development of modern-day international environmental law. The conference marked the worldwide recognition of the environment crises as a matter of international concern, requiring an integrated international response. The conference adopted a Declaration on the Human Environment containing principles, which, although not binding by themselves, subsequently inspired or explicitly found their way into a large number of binding international instruments.

Two decades later, the 1992 UN Conference on Environment and Development (UNCED), held in Rio de Janeiro, constituted the second major milestone in the development of international environmental law. This conference focused on the linkage between the environment and development and the need for sustainable development, as set out in the 1987 Brundtland Report. This approach was reflected in the Rio Declaration on Environment and Development, containing 26 principles, which may be regarded as the successor to the Stockholm Declaration.

One important outcome of the UNCED was the establishment in 1993 of the UN Commission on Sustainable Development. The Commission consists of 53 Member States elected for a period of 3 years. It reports to the UN Economic and Social Council (ECOSOC) and its mandate is to monitor the implementation of Agenda 21 (UNCED's programme action).

These international conferences stimulated the development of international environmental law.

Some of the basic principles of international environmental law are as follows:

1. Environmental protection principles that include the sustainability principle, the polluter pay principle and precautionary principle

2. Social principles that include the equity principle, human rights principle and participation principle

(i) The polluter pays principle: According to this principle, the polluter should bear the expenses of carrying out the anti- pollution measures as decided by the public authorities. The costs of these measures would, thus, be reflected in the costs of goods and services, which cause the pollution.

(ii) Principle of non-discrimination: According to this principle, polluters causing transboundary pollution would be treated no less severely, than they would be, if they caused similar pollution within their own country. This right of equal access entails that a victim of transboundary pollution is granted no less favourable treatments than victims in the country where the pollution originated.

(iii) Precautionary principle: According to this, lack of full scientific certainty shall not be used as a reason for postponing measures to prevent environmental degradation. The actual application of this principle is the measures taken under the 'Montreal Protocol on Substances that Deplete the Ozone Layer'.

(iv) Principle of common differentiated responsibilities: According to this, nations should divide the costs of measures to protect the environment on the basis of the fact that they have made different contributions to global environmental degradation. One of the clearest examples of the application of this principle may be found in the Convention on Climate Change. This convention not only accepts that developing countries need to comply with less strict standards than developed countries; it also accepts that they are entitled to technological and financial assistance in order to help them meet their obligations under the treaty.

(v) Principle of intergenerational equity: According to this, nations are obliged to take into account the long-term effects of their actions affecting the environment. This principle attempts to emphasise that attention should not only be paid to long-distance effects but also to the long-term effects of human activity. After all, present-day decisions may restrict future uses of natural resources and may force upon future generations considerable cleaning-costs. It cannot simply be assumed that future generations will be able to develop the necessary technology for this purpose.

vi) The sustainability principle: According to this principle

The first wave of modern environmentalism was associated with the counter culture movement of the 1960's and 1970s. It has emerged as an awareness of the potential for a global ecological crisis from traditional nature conservation concern. This also introduced world to the concept of 'sustainability', of systems in equilibrium.

Environmentalists argued that the industrial activity and exponential growth of populations could be sustained only by depleting Earth's resources and overloading the planets ability to deal with pollution and waste materials. Between 1965 and 1970 there was a drastic change in public regarding environmental protection. There was a sense of urgency suddenly pervaded public discussion of environmental issues. The press was filled with the stories of environmental trauma. (Vogel 1989:65)

Indian Environmental Laws

Indian environmental laws follow the international law principles but there are few exceptions in accordance with situations and environmental issues in the country.

There has been an urgent need in India to reform various sectors of law and torts that were ignored, mostly for the reasons of high costs of tort litigation. The law of torts (common law) as administered in India in modern times is the English law as found suitable to Indian conditions and as modified by Acts of the Indian Legislature. Certainly, some of the features of the law of torts developed in England are absent in India. The Indian courts therefore, apply those principles to match the situations in India. This means there is an altercation from the British law to suit the Indian conditions. This fact is quite appreciated because there is a difference in the societies and systems of Britain and India. Tort law is not codified in India. This means whenever an exigency arises, a precedent shall be set up to take care of the peculiar situation. This definitely is a good solution to cope up with civil matters in the largest democracy on the planet. It has also been noted in the Union Carbide Case which enables a Civil Court to try all suits of a civil nature, impliedly confers jurisdiction to apply the law of torts as principles of justice, equity and good conscience. This definitely is, providing a forum to try civil wrongs and making them a miscellaneous category. The development of tort law is evident in the law relating to nuisance as well.

Principles, such as the polluter pays principle are now being accepted through various judicial pronouncements in India.

Effective way of controlling pollution and degradation of resources is to combine traditional laws with modern legislation. To protect the limited resources in the country there is a need for it to implement environmental laws that act as safeguards. In India, The Ministry of Environment and Forest (MOEF) is the nodal agency at the central level for planning, promoting and coordinating the environmental programmes, apart from policy formulation. In executing the responsibilities of this agency, number of other enforcement parties assist.

Industries play vital role in the economic development of any country and hence it is also important to manage pollution contributed by industries. In India, the Central Pollution Control Board monitors the industrial pollution at central level and state of departments of Environment and state at state level. These are also statutory authorities attached to MOEF.

Different laws evolved from the modern legisltations of India are discussed, some of the cases are mentioned in this Case 2.2 which only illustrates the judicial activism in the field of environmental jurisprudence by the Supreme Court of India.

Given below is the list of principles laid down by United Nations Conference on Human Environment.

Principle 1 : Man has the fundamental right to freedom, equality and adequate conditions of life, in an environment of a quality that permits a life of dignity and well-being, and he bears a solemn responsibility to protect and improve the environment for present and future generations. In this respect, policies promoting or perpetuating apartheid, racial segregation, discrimination, colonial and other forms of oppression and foreign domination stand condemned and must be eliminated.

Principle 2: The natural resources of the earth, including the air, water, land, flora and fauna and especially representative samples of natural ecosystems, must be safeguarded for the benefit of present and future generations through careful planning or management, as appropriate.

Principle 3: The capacity of the earth to produce vital renewable resources must be maintained and, wherever practicable, restored or improved.

Principle 4: Man has a special responsibility to safeguard and wisely manage the heritage of wildlife and its habitat, which are now gravely imperiled by a combination of adverse factors. Nature conservation, including wildlife, must therefore receive importance in planning for economic development.

Principle 5 : The non-renewable resources of the earth must be employed in such a way as to guard against the danger of their future exhaustion and to ensure that benefits from such employment are shared by all mankind.

Principle 6 : The discharge of toxic substances or of other substances and the release of heat, in such quantities or concentrations as to exceed the capacity of the environment to render them harmless, must be halted in order to ensure that serious or irreversible damage is not inflicted upon ecosystems. The just struggle of the peoples of ill countries against pollution should be supported.

Principle 7 : States shall take all possible steps to prevent pollution of the seas by substances that are liable to create hazards to human health, to harm living resources and marine life, to damage amenities or to interfere with other legitimate uses of the sea.

Principle 8 : Economic and social development is essential for ensuring a favorable living and working environment for man and for creating conditions on earth that are necessary for the improvement of the quality of life.

Principle 9: En vironmental deficiencies generated by the conditions of under- development and natural disasters pose grave problems and can best be remedied by accelerated development through the transfer of substantial quantities of financial and technological assistance as a supplement to the domestic effort of the developing countries and such timely assistance as may be required.

Principle 10: For the developing countries, stability of prices and adequate earnings for primary commodities and raw materials are essential to environmental management, since economic factors as well as ecological processes must be taken into account.

Principle 11: The environmental policies of all States should enhance and not adversely affect the present or future development potential of developing countries, nor should they hamper the attainment of better living conditions for all, and appropriate steps should be taken by States and international organizations with a view to reaching agreement on meeting the possible national and international economic consequences resulting from the application of environmental measures.

Principle 12: Resources should be made available to preserve and improve the environment, taking into account the circumstances and particular requirements of developing countries and any costs which may emanate- from their incorporating environmental safeguards into their development planning and the need for making available to them, upon their request, additional international technical and financial assistance for this purpose.

Principle 13: In order to achieve a more rational management of resources and thus to improve the environment, States should adopt an integrated and coordinated approach to their development planning so as to ensure that development is compatible with the need to protect and improve environment for the benefit of their population.

Principle 14: Rational planning constitutes an essential tool for reconciling any conflict between the needs of development and the need to protect and improve the environment.

Principle 15: Planning must be applied to human settlements and urbanization with a view to avoiding adverse effects on the environment and obtaining maximum social, economic and environmental benefits for all. In this respect projects which are designed for colonialist and racist domination must be abandoned.

Principle 16: Demographic policies which are without prejudice to basic human rights and which are deemed appropriate by Governm ents concerned should be applied in those regions where the rate of population growth or excessive population concentrations are likely to have adverse effects on the environment of the human environment and impede development.

Principle 17: Appropriate national institutions must be entrusted with the task of planning, managing or controlling the 9 environmental resources of States with a view to enhancing environmental quality.

Principle 18: Science and technology, as part of their contribution to economic and social development, must be applied to the identification, avoidance and control of environmental risks and the solution of environmental problems and for the common good of mankind.

Principle 19: Education in environmental matters, for the younger generation as well as adults, giving due consideration to the underprivileged, is essential in order to broaden the basis for an enlightened opinion and responsible conduct by individuals, enterprises and communities in protecting and improving the environme nt in its full human dimension. It is also essential that mass media of communications avoid contributing to the deterioration of the environment, but, on the contrary, disseminates information of an educational nature on the need to project and improve the environment in order to enable mal to develop in every respect.

Principle 20: Scientific research and development in the context of environmental problems, both national and multinational, must be promoted in all countries, especially the developing countries. In this connection, the free flow of up-to-date scientific information and transfer of experience must be supported and assisted, to facilitate the solution of environmental problems; environmental technologies should be made available to developing countries on terms which would encourage their wide dissemination without constituting an economic burden on the developing countries.

Principle 21: States have, in accordance with the Charter of the United Nations and the principles of international law, the sovereign right to exploit their own resources pursuant to their own environmental policies, and the responsibility to ensure that activities within their jurisdiction or control do not cause damage to the environment of other States or of areas beyond the limits of national jurisdiction.

Principle 22: States shall cooperate to develop further the international law regarding liability and compensation for the victims of pollution and other environmental damage caused by activities within the jurisdicti on or control of such States to areas beyond their jurisdiction.

Principle 23: Without prejudice to such criteria as may be agreed upon by the international community, or to standards which will have to be determined nationally, it will be essential in all cases to consider the systems of values prevailing in each country, and the extent of the applicability of standards which are valid for the most advanced countries but which may be inappropriate and of unwarranted social cost for the developing countries.

Principle 24: International matters concerning the protection and improvement of the environment should be handled in a cooperative spirit by all countries, big and small, on an equal footing. Cooperation through multilateral or bilateral arrangements or other appropriate means is essential to effectively control, prevent, reduce and eliminate adverse environmental effects resulting from activities conducted in all spheres; in such a way that due account is taken of the sovereignty and interests of all States.

Principle 25: States shall ensure that international organizations play a coordinated, efficient and dynamic role for the protection and improvement of the environment.

Principle 26: Man and his environment must be spared the effects of nuclear weapons and all other means of mass destruction. States must strive to reach prompt agreement, in the relevant international organs, on the elimination and complete destruction of such weapons.

International Institutions

Now, let us touch upon some of the international institutions and agencies that were instrumental in bringing about environmental reforms through laws and legislations. These include:

- The United Nations Environment Programme (UNEP): Set up in the wake of the Stockholm

Conference, UNEP is not a specialised agency but merely a 'programme', without executive powers and with a budget to a large extent dependent on voluntary contributions from States. UNEP's main programme consists of better co-ordination between existing UN programmes in the field of the environment, rather than to develop its own programmes. Nevertheless, UNEP has gradually managed to carve out a useful role for itself by taking the initiative on such issues as the control of transboundary movements of hazardous waste and the protection of the ozone layer.

- The Food and Agricultural Organisation (FAO): This has been active on subjects such as deforestation and conservation of fisheries.

- The World Meteorological Organisation (WMO): This plays a crucial role in the field of climate change and global warming.

- The International Atomic Energy Agency (IAEA): This agency although not initially much concerned on the subject, but in the wake of the Chernobyl disaster in Russia has focused its attention towards environmental pollution.

- International Bank for Reconstruction and Development (IBRD)/The World Bank (WB): This has, in recent years, begun to pay more attention to the environmental side effects of its lending policies.

Outside the UN system, regional organisations, which have adopted significant legal instruments in the field of the environment, include the Organisation for Economic Co-operation and Development (OECD), the Conference on Security and Cooperation in Europe (CSCE) and the Organisation of African Unity (OAU).

Key International Treaties

The following are the important international environmental treaties that deal with air, water, hazardous waste, nuclear wastes, etc.:

(i) The 1992 Helsinki Convention on the protection and use of trans-boundary watercourses and international lakes. (This convention deals with international watercourses.)

(ii) The 1973 Convention for the prevention of pollution from ships (MARPOL).

(iii) The 1982 UN Convention on the law of the sea (UNCLOS).

(iv) The 1992 Paris Convention for the protection of the marine environment of the North-East Atlantic. (The conventions listed at (ii) to (iv) deal with marine pollution.)

(v) The 1979 ECE Convention on long-range trans-boundary air pollution-protocols on sulphur dioxide (SO_2), nitrogen oxides (NOX) and volatile organic compounds (VOCs).

(vi) The 1985 Vienna Convention for the protection of the ozone layer (Montreal Protocol and London and Copenhagen Amendments).

(vii) The 1992 UN Convention on climate change. (The conventions listed at (v) to (vii) deal with atmospheric pollution.)

(viii) The 1989 Basel Convention on the control of trans-boundary movements of hazardous wastes and their disposal.

(ix) The 1991 Bamako Convention on the ban of the import into Africa and the control of trans-boundary movement and management of hazardous wastes within Africa. (The conventions listed at (viii) and (ix) deal with hazardous waste.)

(x) The 1986 IAEA Convention on early notification of a nuclear accident.

(xi) The 1986 IAEA Convention on assistance in the case of a nuclear accident.

(xii) 1994 IAEA Convention on nuclear safety. (The conventions listed at (x) to (xii) deal with nuclear energy.)

(xiii) The 1991 ECE Convention on environmental impact assessment in a trans-boundary context. (xiv) The 1992 UN Convention on biological diversity.

Objectives and Principles of Legislation

The objectives of environmental legislation are to provide a set of enforceable and standard rules to contribute to the pursuit of:

(i) preserving, protecting and improving the quality of the environment;

(ii) protecting human health;

(iii) utilising natural resources in a prudent and rational way;

(iv) promoting measures at international level to deal with regional or worldwide environmental problems.

Environmental legislation seeks to regulate pollution of the natural environment in relation to air, noise, vibration, water, radiation and soil. It contains rules relating to the conservation of the natural environment, the protection of endangered species, the promotion of biological diversity, the protection of forests and the pursuit of environmentally friendly agriculture. With regard to the human environment, it seeks to protect human beings (the consumer) against contaminated food, dangerous or defective products, economic harm and danger in travel. With regard to the humanmade environment it seeks the protection of historic and cultural environment. It seeks harmonisation of standards and the enforcement of those standards through legislative provisions particularly relating to the introduction of appropriate environmental management systems to harness scarce resources.

The further objectives of environmental legislation are to set standards such as achieving a high level of protection by taking into account the diversity of situations in the various regions, to give a legal basis to the precautionary principle whereby legislative action is taken where there is no reason to believe that substances or energy or materials introduced directly or indirectly into the environment, may, or are likely to create, hazards to human health, harm living resources, damage communities or interfere with other legitimate uses. This can be done even where there is no conclusive evidence of a cause or relationship between inputs and their effects and to give a legal basis to the principle that preventive action should be taken as prevention and if successful, to advise all the detailed legislation relating to steps to be taken to cure a problem.

Environmental legislation is generally based on the principles that environmental damage should, as a priority, be rectified at source and that the polluter should pay. Environmental protection requirements should be integrated into the definition and implementation of legislation relating to non-environmental issues. Where cross border matters relating to provisions primarily of a fiscal matter, measures concerning town and country planning and land use (other than waste management), management of water resources and measures significantly affecting the choice between different energy resources and the general structure of energy supply, decisions should only be adopted with unanimous agreement between the participants.

- Other legal principles produced by the expert group are that countries must:

- conserve and use the environment including its natural resources for the benefit of both the present and future generations;

- maintain ecosystems and ecological processes essential for the functioning of the biosphere, and preserve biological diversity;

- observe the principle of optimum sustainable yield in use when dealing with natural resources and ecosystems;

- establish adequate environmental protection standards and monitor changes and publish relevant data on environmental quality and resource use.

Environmental Legislations in India

India has had a long history of environmentalism with the passage and codification of acts such as the Indian Penal Code, The Criminal Procedure Code, The Bengal Smoke Nuisance Act of 1905, The Indian Motor Vehicle Act, The Factories Act, The Indian Forest Act, The Mines and Minerals (Regulation and Development) Act, The Industries (Development and Regulation) Act, The Forest (Conservation) Act and The Merchant Shipping Act.

The Indian Penal Code, passed in 1860, penalises persons responsible for causing defilement of water of a public spring or reservoir with imprisonment or fines. Traditionally, the interpretation of the Code has been viewed as a conservative attempt at enforcement. This is because punishment and fines have been characterised as meagre. In addition, fouling a "public spring" has not, by definition, also included a "public river", which is where most pollution occurs. Finally, the specific language of the code places the burden of proof on the prosecution. Polluters must "voluntarily", "with intent", or "knowingly" discharge damaging effluents, making successful prosecution problematic in a court of law.

The Factories Act also addresses public safety and health issues. The legislation addresses the discharge of water and effluents by factories, calling for effective arrangements for disposal at the plant-level. As in the Indian Penal Code, penalties have been provided.

Section 12 of the Act empowered each State government to legislate its own rules. As a result, a number of states passed versions of the Factory Act, including Uttar Pradesh in 1950, Tamil Nadu in 1950, West Bengal in 1958, Maharashtra in 1963 and Mysore in 1969. Each tailored the Act to suit its particular situation. In Uttar Pradesh, for example, disposal of effluents had to have the approval of the State's pollution board. In Tamil Nadu, the ruling entity with similar responsibilities was the Director of Fisheries. In Maharashtra, local authorities were granted with jurisdiction in such matters.

The watershed event in the environmental movement was the Stockholm Conference on Human Environment in June 1972. The conference made it apparent to all attendees that each nation needed to adopt comprehensive legislation addressing health and safety issues for people, flora and fauna. The United Nations, organisers of the conference, requested each participant to provide a country report. Stockholm, thus, served as the genesis for the series of environmental measures India passed in the years to come.

Evolution of Environmental Legislation in India

The Indian Forest Act, 1927 consolidates the law relating to forests, the transit of forest-produce and the duty leviable on timber and other forest products.

The Prevention of Cruelty to Animals Act was enacted in 1960 to prevent the infliction of unnecessary suffering on the animals and to amend the laws relating to the prevention of cruelty to animals. As a promotion for enactment of this act there was formation of animal board of India.

n 1966 Indian Forest Service was constituted under the All India Services Act, 1951 by the government of India. The main aim of their service is to implement the country's National Forest Policy which envisages scientific management of forest and to exploit them on a sustained basis primarily for timber products.

Wild life Act enacted in the year 1972 with the objective of effectively protecting the wild life of the country and to control poaching, smuggling and illegal trade in wildlife and its derivatives. This act was amended in January 2003. To strengthen the act; the Ministry has proposed further amendments in the law by introducing more rigid measures. Main objective is to provide protection to the flora and fauna and also to ecologically important protected areas.

Water Act was enacted in 1974 to provide for the prevention and control of water pollution and for water maintenance in the country. The Water cess Act was enacted in 1977, to provide for the levy and collection of a cess on water consumed by persons operating and carrying on certain types of industrial activities. The act was last amended in 2003.

Forest Conservation Act was enacted in 1980 to protect and conserve country's forest.

Air Act (1981) and was amended in 1987 to provide for prevention control and abatement of air pollution in India.

Well know Environment protection Act (1986) came into existence after 14 years of UN conference with an objective of protection and improvement of the country. Later on the amendments were done to it in 1991.

The Man and Biosphere (MAB) programme of UNESCO was launched in 1971; India joined it in 1988 after formation of bioreserve committee. Purpose of this is to develop a base for rational use or conservation of natural resources while improving the relationship between the man and environment.

In India the coast line is very lengthy which runs to 7860 km. The coastal line consists of Malvan (Maharashtra), Okha (Gujarat), Mandapam (Tamil Nadu), Gangetic Sundarbans (West Bengal) as well as Lakshadweep and Andaman group of islands which are rich with regard to the marine flora and fauna. When we are blessed with the natural resources, it is our duty to safeguard it and pass it on to our posterity. India has now established 15 bioreserves the first one Nokrek (Meghalaya) in 1988.

In Hazardous waste rules (1989) were framed in which hazardous chemicals list was finalised.

The Eco-Mark Scheme of India was introduced in 1991 to increase the environmental awareness amongst citizens. This scheme aimed at encouraging the public to purchase products which are eco friendly.

Public liability insurance act was enacted in 1991 to provide for damages to victims of an accident which occurs as a result of handling hazardous substances (owners associated with the production or handling).

National Environment Tribunal (1995) is for strict liability for damage arising out of accidents caused from handling of hazardous waste.

Biomedical Waste Rules (1998) that deal with collection, reception, storage, treatment and disposal of the waste.

The Noise Pollution Rules (2000) the state government categorised industrial, commercial and residential or silence zones to implement noise standards.

The Biodiversity Act (2002) was born out of India's attempt to realise the objectives mentioned in the United Nations convention on biological Diversity (CBD) enacted in 1992 states that country should use their own biological resources.

The Scheduled Tribes and Other Traditional Forest Dwellers (Recognition of Forest Rights) Act, 2006, recognizes the rights of forest-dwelling Scheduled Tribes and other traditional forest dwellers over the forest areas inhabited by them. This act also provides framework for their rights.

The National Environment Appellate Authority (NEAA) was set up by the ministry of environment and forests to address cases in which environment clearances are required in certain restricted areas. It was established by the National Environment Appellate Authority Act 1997 to hear appeals with respect to restriction of areas in which any industries, operations or processes, operations or processes shall or shall not be carried out, subject to certain safeguards under the Environment (Protection) Act, 1986. The Authority shall become defunct and the Act shall stand repealed upon the enactment of the National Green Tribunal Bill 2009 currently pending in Parliament.

Constitution of India

India was the first country to insert an amendment into its Constitution allowing the State to protect and improve the environment for safeguarding public health, forests and wild life. The 42nd amendment was adopted in 1976 and went into effect on 3 rd January 1977. The language of the Directive Principles of State Policy (Article 47) requires not only a protectionist stance by the State but also compels the State to seek the improvement of polluted environments. This allows the government to impose restrictions on potentially harmful entities such as polluting industries.

An important subtlety of the directive's language is the provision that the article "shall not be enforceable by any court, but it shall be the duty of the State to apply these principles in making laws." This allows the directive to be an instrument of guidance for the government, while at the same time, since no law has been passed, no individual can violate existing law.

Although State governments have clearly delineated lines of authority and jurisdiction, Article 253 of the Constitution provides the Union government with sweeping powers to implement laws for any part of India with regard to treaties made with another country or decisions made by an international body. For internal environmental matters, the Constitution provides for a distribution of legislative powers between the Union and the States. This was done by the creation of three jurisdictional listings – Union, State and Concurrent.

The central government can enact a law on any items on the Union and Concurrent lists, and, in certain cases, may enact a law on an item on the State list. To do so, the central government must approach the state legislatures. In the case of "water", which is an entry on the State list, Article 252 requires at least two or more State legislatures pass resolutions empowering the Parliament to pass water-related legislation. Approaching the State on issues within its jurisdiction has historically led to cumbersome delays. For example, it was almost a decade after the first draft bill was being debated, before the Parliament was able to pass India's first major water legislation in 1974. The process of State legislatures voting to empower the Parliament to pass water-related measures repeated itself during subsequent amendments to the water law in 1978 and 1988.

Union Government Initiatives

As early as 1962, the Union Ministry of Health had begun to address water pollution issues by appointing a study committee. The committee made recommendations for both central and state level action. Jurisdictional questions remained unsolved between the States and Union government, but by 1965 a draft bill was finally being circulated which allowed the States to pass resolutions authorising Parliament to enact legislation on their behalf. By 1969, a bill, the Prevention of Water Pollution, had been introduced. Ultimately, a modified version, i.e., the Water (Prevention and Control of Pollution) Act, was passed in 1974. Institutionalising a regulatory agency for controlling water pollution marked the first true commitment on the issue by the Indian Parliament. The Water Act also established the Pollution Control Boards at central government and state government levels.

n the aftermath of the Water Act, the then Prime Minister Indira Gandhi moved to enact a series of environmental measures. The Department of Environment (DOE), created in 1980, performed an oversight role for the Union government. DOE did environmental appraisals of development projects, monitored air and water quality, established an environmental information system, promoted research, and coordinated activities between Union, State and local governments.

Establishment of Ministry of Environment & Forest (MOEF)

The government of the then Indian Prime Minister Rajiv Gandhi recognised the deficiencies and created in 1985 the Ministry of Environment and Forests (MOEF). MOEF was more comprehensive and institutionalised, and had a Union Minister and Minister of State, two political positions answering directly to the Prime Minister. The agency comprised of eighteen divisions and two independent units, the Ganga Project Directorate and the National Mission on Wastelands Development. It continued the same functions that DOE originally had, such as monitoring and enforcement, conducting environmental assessments and surveys, but also did promotional work about the environment. In 1981, the Air (Prevention and Control of Pollution) Act was passed, and in 1986, the Parliament passed the Environment (Protection) Act, designed to act as umbrella legislation on the environment. The responsibility entrusted to administer the new legislation also fell to the Union and State pollution control boards.

In December 1993, the MOEF completed its Environmental Action Plan to integrate environmental considerations into developmental strategies, which, among other priorities, included industrial pollution reduction. In principle, though, the MOEF was already actively pursuing pollution abatement and prevention policies. Among the strategies employed by the MOEF was the implementation of a polluter pays principle. MOEF imposed significant increases in its water cess, increasing it six-fold from early 1992. Since then, the MOEF has decided to increase the water cess again since the charges were still well below the current effluent treatment costs.

The MOEF also decided to shift from concentration to load-based standards. This would add to a polluter's costs and remove incentives to dilute effluents by adding water, and strengthen incentives for adoption of cleaner technologies. It also issued water consumption standards, which were an additional charge for excessive water use.

Targeting small-scale industries has been an important task as well, since these facilities greatly add to the pollution load. The Ministry provides technical assistance and limited grants to promote central effluent treatment plants. It has also created industrial zones to encourage clusters of similar industries in order to help reduce the cost of providing utilities and environmental services.

In January 1994, the Union government introduced new restrictions and prohibitions on the expansion and/or modernisation of any activity or new project, unless an environmental clearance is granted. Projects must, for example, submit an environmental impact assessment (EIA) and environmental management plan (EMP). The industry sector and the size of the project determine whether the MOEF or the State government has jurisdiction in assessing the EIA. However, in some highly polluting sectors such as power plants, cement and petrochemicals, the MOEF generally clears the EIA. In addition, State governments may issue more stringent restrictions for heavily polluted areas. One example is the Vapi region in Gujarat, where the State now has a total ban on new activities for the country's 17 most highly polluting sectors.

The Ministry has issued the Environmental Impact Assessment Notification, 2006, which makes environmental clearance mandatory for the development activities listed in its schedule.

References

- Alan P. Loeb, "Addressing the Public's Goals for Environmental Regulation When Communicating Acid Rain Allowance Trades," The Electricity Journal, May 1995
- Monterey Institute of International Studies. "MA in International Environmental Policy". Miis.edu. Retrieved 2012-11-02

- Rushefsky, Mark E. (2002). Public Policy in the United States at the Dawn of the Twenty-first Century (3rd ed.). New York: M.E. Sharpe, Inc. pp. 253–254. ISBN 978-0-7656-1663-0

- Major article outlining and analyzing the history of environmental communication policy within the European Union has recently come out in The Information Society, a journal based in the United States

- Eccleston, Charles H. (2010). Global Environmental Policy: Concepts, Principles, and Practice. Chapter 7. ISBN 978-1439847664

4

System Standards Related to Environmental Management

Environmental Management System refers to managing an organization to achieve its environmental goals efficiently. The most widely recognized standards that the system is based on are ISO 14000 and ISO 14001. The chapter also discusses the Life Cycle Assessment (LCA), which is a technique that evaluates the environmental impact a product cause throughout its entire life cycle. The major categories of Environmental Management System Standards and Life Cycle Assessment are dealt with great details in the chapter.

Environmental Management System

Environmental management system (EMS) refers to the management of an organization's environmental programs in a comprehensive, systematic, planned and documented manner. It includes the organizational structure, planning and resources for developing, implementing and maintaining policy for environmental protection.

More formally, EMS is "a system and database which integrates procedures and processes for training of personnel, monitoring, summarizing, and reporting of specialized environmental performance information to internal and external stakeholders of a firm."

The most widely used standard on which an EMS is based is International Organization for Standardization (ISO) 14001. Alternatives include the EMAS.

An environmental management information system (EMIS) is an information technology solution for tracking environmental data for a company as part of their overall environmental management system.

Goals

The goals of EMS are to increase compliance and reduce waste:

- Compliance is the act of reaching and maintaining minimal legal standards. By not being compliant, companies may face fines, government intervention or may not be able to operate.

- Waste reduction goes beyond compliance to reduce environmental impact. The EMS helps to develop, implement, manage, coordinate and monitor environmental policies. Waste reduction begins at the design phase through pollution prevention and waste minimization. At the end of the life cycle, waste is reduced by recycling.

To meet these goals, the selection of environmental management systems is typically subject to a certain set of criteria: a proven capability to handle high frequency data, high performance indicators, transparent handling and processing of data, powerful calculation engine, customised factor handling, multiple integration capabilities, automation of workflows and QA processes and in-depth, flexible reporting.

Features

An environmental management system (EMS):

- Serves as a tool, or process, to improve environmental performance and information mainly "design, pollution control and waste minimization, training, reporting to top management, and the setting of goals"

- Provides a systematic way of managing an organization's environmental affairs

- Is the aspect of the organization's overall management structure that addresses immediate and long-term impacts of its products, services and processes on the environment. EMS assists with planning, controlling and monitoring policies in an organization.

- Gives order and consistency for organizations to address environmental concerns through the allocation of resources, assignment of responsibility and ongoing evaluation of practices, procedures and processes

- Creates environmental buy-in from management and employees and assigns accountability and responsibility.

- Sets framework for training to achieve objectives and desired performance.

- Helps understand legislative requirements to better determine a product or service's impact, significance, priorities and objectives.

- Focuses on continual improvement of the system and a way to implement policies and objectives to meet a desired result. This also helps with reviewing and auditing the EMS to find future opportunities.

- Encourages contractors and suppliers to establish their own EMS.

EMS Model

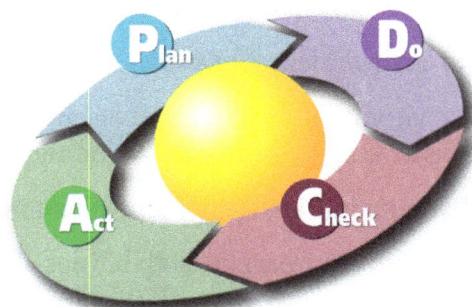

The PDCA cycle

An EMS follows a Plan-Do-Check-Act, or PDCA, Cycle. The diagram shows the process of first developing an environmental policy, planning the EMS, and then implementing it. The process also includes checking the system and acting on it. The model is continuous because an EMS is a process of continual improvement in which an organization is constantly reviewing and revising the system.

This is a model that can be used by a wide range of organizations — from manufacturing facilities to service industries to government agencies.

Other Meanings

An EMS can also be classified as:

- a system which monitors, tracks and reports emissions information, particularly with respect to the oil and gas industry. EMSs are becoming web-based in response to the EPA's mandated greenhouse gas (GHG) reporting rule, which allows for reporting GHG emissions information via the internet.

- a centrally controlled and often automated network of devices (now frequently wireless using z-wave and zigbee technologies) used to control the internal environment of a building. Such a system namely acts as an interface between end user and energy (gas/electricity) consumption.

Examples of Environmental Management Systems

- Emisoft's environmental management, reporting and compliance platform

- Medgate environmental management software

- EsDat environmental data management system

- Enviance regulatory compliance system

The growing public interest in environmental issues and concern for environmental quality has led to the emergence of strict pollution control regimes. This has brought about the development and implementation of various voluntary schemes and standards for environmental management and pollution control. Notable examples are BS 7750, the European Union's Eco-Management and Audit Scheme (EMAS) and the ISO 14000 series. In this context, putting in place systems to meet with these standards assume significance. An environmental management system (EMS) of an organisation is a system designed to:

- meet the regulatory and legislative system requirements;

- improve the control of the environmental impact;

- provide confidence to the customers that the products and services are manufactured with the aim of reducing the negative impact on the environment;

- suitably accommodate changing market trends and gain competitive edge;

- reduce the costs associated with environmental liabilities and insurance;

- gain public and media support.

EMS involves preparing a list of all environmental legislation, making a register of all environmental effects associated with the company's activities, setting targets (such as reducing waste production by a quantifiable amount), keeping appropriate records and undertaking regular reviews of the system. The whole system, in essence, forms the basis for sound environmental performance.

Most EMSs are built on the notational model. This 'plan-do-check-act' model leads to continuous improvement based upon:

- planning, including identifying environmental aspects and establishing goals (plan);

- implementing, including training and operational controls (do);

- checking, including monitoring and corrective action (check);

- reviewing, including progress reviews and acting to incorporate required changes in the EMS (act).

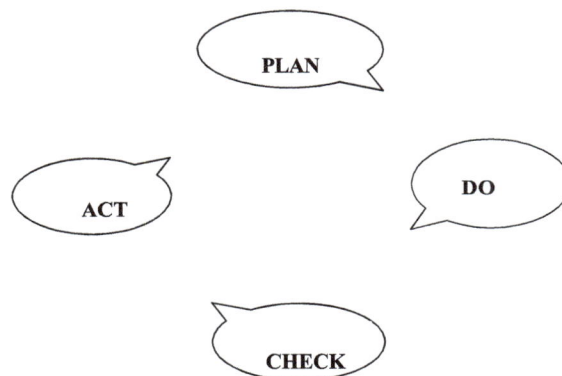

The Plan – Do – Check – Act Model

To reiterate, this model ensures that environmental matters are systematically identified, controlled and monitored. It also helps ensure that performance of the EMS is improved over time.

Core Elements of EMS

There is no single best approach to the development and implementation of EMS, since it depends on the nature, size and complexity of the activities, products and services within the organisation. All environmental management systems, however, have a number of core elements in common and these include the following:.

- Environmental policy: An environmental policy is usually published as a written statement, expressing the commitment of the senior management to improving appropriate environmental performance. It is most often understood as a public statement of the intentions and principles of action for the organisation regarding the environment. The policy statement should define the broad environmental goals the organisation has decided to achieve. These are most clear if they are quantified, e.g., to reduce emissions of pollutants by 95% within 5 years, to provide sewerage and biological treatment of sludge for 60% of the population within 3 years, etc.

- Environment action plan: An environment programme or action plan describes the measures the organisation will take over the coming year(s). The action plan essentially translates the environmental policies of the organisation into objectives and targets and identifies the activities to achieve them, defines responsibilities and commits the necessary human and financial resources for implementation. This includes committing the necessary funds and staff to meet each goal and providing for monitoring and co-ordination of the progress made towards achieving these separate goals and the overall policy goals that have to be fulfilled.

- Environment related organisational structures: The structures establish assignments, delegate authority and assign responsibility for actions. In the case of organisations with multiple sites or different activities, structures for the organisation as a whole as well as for the separate operating units are necessary. The senior staff member responsible for the environment, typically, has a direct reporting relationship to the head of the organisation. Individuals holding strategic or main environmental responsibilities should be adequately supported with human and financial resources.

- Integration of environmental concerns: The integration of environmental management into regular operation means the establishment of procedures for incorporating environmental measures into other operational aspects such as the protection of workers, purchasing, R&D, product development and acquisitions, marketing, finance, etc., in the case of companies and the safety, health and welfare of the community in the case of a local government. This encompasses the development of specific environmental procedures, usually detailed in operating manuals and other operating instructions describing measures and actions to take in the implementation of the environmental programme or action plan.

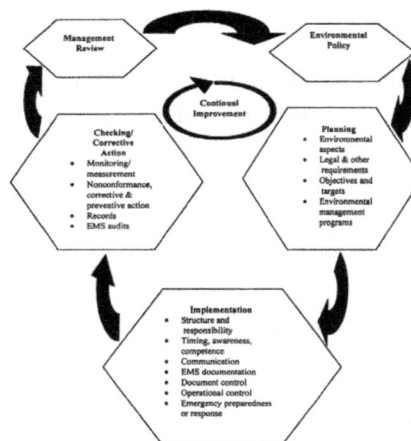

Because of the cyclic nature of the model, EMS is very dynamic in that a change or revision in any of the core elements of the system will have a chain/cascading effect on other elements.

Note that while there are several EMS models available, the model presented above uses the standards set by ISO 14001 as a starting point for describing EMS elements. This is so, mainly, due to the fact that:

- ISO 14001 is a widely accepted international standard for EMS that focuses on continual improvement.

- Companies may be asked to demonstrate conformance with ISO 14001 as a condition of doing business in some markets.

Benefits of EMS

A number of benefits are associated with adoption of EMS. For example, insurance companies are beginning to exclude pollution incidents from their insurance cover, and when they are included, the insurance companies place a surcharge on the policies. In order to improve or reduce any increases in the insurance premiums, companies can carry out an environmental audit and implement a EMS.

From a managerial point of view, a EMS enables a company to develop and maintain a well-organised management structure that ensures compliance with environmental legislation. By keeping the working place cleaner and safer, and the production more efficient, staff interest can be aroused and morale improved. Another incentive is the potential financial gains, which can be achieved. Reducing wastes and emissions and improving their treatment helps many companies to save a considerable sum of money in the long term. Adopting a EMS will also facilitate trade, as major trading blocs are more likely to accept members with a certain level of environmental performance. In other words, the benefits of a EMS include:

- Improved environmental performance.

- Enhanced compliance.

- Prevention of pollution/resource conservation.

- New customers/markets.

- Increased efficiency/reduced costs.

- Enhanced employee morale.

- Enhanced image with public, regulators, lenders, investors.

- Employee awareness of environmental issues and responsibilities.

Certification Body Assessments of EMS

To be certified, an organisation has to demonstrate that the EMS functions and the various control mechanisms are properly operational. In practice, this means that the EMS has been operational for a minimum of 3 months, the internal audit system is fully operational and that one management review has been conducted (Tech Monitor, Vol. 12. No. 5 Sept-Oct 1995).

Based on the following conditions, certificates are issued:

Non-conformities identified have been addressed and eliminated.

The certification body has justified confidence that all provisions in the EMS standard have been met, and, in particular, that provision for compliance with the organisation's policy and objectives is effective.

The principle of continuous improvement of environmental performance has been stated within an environmental programme and is being adhered to play in the EMS.

All key staff (i.e., those involved in managing significant effects) have undergone a training need analysis and have received training.

Regulatory Compliance

An organisation with a certified EMS has a system that should achieve continuing compliance with regulatory requirements applicable to the environmental aspects of its activities, products and services. The certification body confirms that a system designed to achieve the necessary compliance is operating effectively. In situations where authorities issue licenses or permits, it will often be a requirement that the organisation notifies the authority of any breaches. The certification body should ensure that the organisation's EMS records any infringement of regulations and that the appropriate corrective action has been taken.

An organisation with a certified EMS is responsible for continuing compliance with all the regulatory requirements and, therefore, must maintain a record of incidents of non-compliance and the remedial action.

Surveillance Audits

At each surveillance, the certification body should pay attention to the following:

- the effectiveness of the EMS with regard to achieving the objectives of the organisation's environmental policy;

- an interview with management responsible for the EMS;

- the functioning of procedures for notifying the authorities of any breaches;

- progress of planned activities aimed at continuous improvement of environmental performance, where applicable;

- follow up conclusions resulting from internal audits;

- action taken on non-conformities identified during the previous audit

The certification body should be able to adapt its surveillance programme to the environmental issues related to the activities of the organisation and justify this arrangement. The surveillance audit of the certification body should be agreed with the organisation, taking into account the internal audit programme and the reliability that can be attributed to this work.

Documentation for EMS

While data relates to fact, information refers to a series of data analysed to provide a decision. Information is an answer to a question. All information comes from data, but not all data come from information (Tech Monitor, Vol. 12. No. 5 Sept-Oct 1995). In other words, executives making decisions often receive excessive amounts of data, but they do not always get the information they need, presented clearly and adequately for use in decisionmaking.

Decisions are usually related to solving problems. In order to solve a problem, one needs information to decide. A EMS, therefore, requires a proper documentation system in order to collect, analyse, register and retrieve information.

The documentation should describe the EMS and make clear the relationship to any other management system in operation within the organisation, or as having an influence on the operations of the entity, subject to certification. It is acceptable to combine the documentation for environmental and other management systems (such as for quality of health and safety), as long as the EMS can be clearly identified, together with the appropriate interfaces between different EMS documentations and the order of precedence.

Environmental effects may have been considered within safety reviews, hazard and operability studies and risk/consequence analysis, etc. In such cases, the EMS documentation should refer to the critical areas where significant effects are covered by other management systems. The EMS documentation should be available to all appropriate staff and should be kept up-to-date.

Environmental Management Manual

A key EMS document is a company's environment manual. This is a document that establishes the general policy of a company on this issue. It usually contains an environmental policy and a clear statement on the person who is in charge of verification of activities in the organisational structure pertaining to the maintenance of the environmental system. It lays down how each requirement of the policy will be handled in the organisation and includes a list of all procedures.

The manual also contains a description of the system, which was or is being implemented. There are considerable benefits in preparing a manual:

- clarifies responsibilities;

- is useful for audit purposes;

- can be highly effective for training of new employees;

- makes easier the process of obtaining approval of licenses from protection agencies;

- is useful for marketing purposes.

The highest executive in the company must sign the document. It should contain the vision of the organisation about environment and its commitment to continuous improvement. The document should detail targets to be reached and explain how they will be reached, with what resources and under whose responsibility. The company must also realise that it is necessary to measure results in order to evaluate how correctly the policy is being applied. The company should, therefore, have a programme for achieving its objectives and targets.

To prepare the manual, it is necessary first to identify all mandatory regulations, standards and codes to be complied with. In addition, the company needs to identify other requirements, such as headquarters' policies, top management's strategic vision, market demands, etc.

The discussions that we have had so far provide you with a basic knowledge of EMS. Now, we will next discuss the importance of ISO 14000 and its supporting systems. Before we do so, let us first do Learning Activity 7.1.

ISO 14000

ISO 14000 is a family of standards related to environmental management that exists to help organizations (a) minimize how their operations (processes, etc.) negatively affect the environment (i.e., cause adverse changes to air, water, or land); (b) comply with applicable laws, regulations, and other environmentally oriented requirements; and (c) continually improve in the above.

The current version of ISO 14001 is ISO 14001:2015 which was published in September 2015.

ISO 14000 is similar to ISO 9000 quality management in that both pertain to the process of how a product is produced, rather than to the product itself. As with ISO 9001, certification is performed by third-party organizations rather than being awarded by ISO directly. The ISO 19011 ISO 17021 audit standard apply when auditins are being performed.

The requirements of ISO 14001 are an integral part of the European Union's Eco-Management and Audit Scheme (EMAS). EMAS's structure and material requirements are more demanding, mainly concerning performance improvement, legal compliance, and reporting duties.

Brief History of Environmental Management Systems

In 1992, BSI Group published the world's first environmental management systems standard, BS 7750. Prior to this, environmental management had been part of larger systems such as Responsible Care. BS 7750 supplied the template for the development of the ISO 14000 series in 1996, by the International Organization for Standardization, which has representation from committees all over the world (ISO) (Clements 1996, Brorson & Larsson, 1999). As of 2010, ISO 14001 is now used by at least 223 149 organizations in 159 countries and economies.

Development of the ISO 14000 Series

The ISO 14000 family includes most notably the ISO 14001 standard, which represents the core set of standards used by organizations for designing and implementing an effective Environmental Management System (EMS). Other standards included in this series are ISO 14004, which gives additional guidelines for a good EMS, and more specialized standards dealing with specific aspects of environmental management. The major objective of the ISO 14000 series of norms is "to promote more effective and efficient environmental management in organizations and to provide useful and usable tools--ones that are cost-effective, system-based, [and] flexible, and reflect the best organizations and the best organizational practices available for gathering, interpreting, and communicating environmentally relevant information".

ISO 14000 series is based on a voluntary approach to environmental regulation (Szymanski & Tiwari 2004). The series includes the ISO 14001 standard, which provides guidelines for the establishment or improvement of an EMS. The standard shares many common traits with its predecessor, ISO 9000, the international standard of quality management (Jackson 1997), which served as a model for its internal structure (National Academy Press 1999), and both can be implemented side by side. As with ISO 9000, ISO 14000 acts both as an internal management tool and as a way of demonstrating a company's environmental commitment to its customers and clients (Boiral 2007).

Prior to the development of the ISO 14000 series, organizations voluntarily constructed their own EMSs, but this made comparisons of environmental effects between companies difficult; therefore, the universal ISO 14000 series was developed. An EMS is defined by ISO as: "part of the overall management system, that includes organizational structure, planning activities, responsibilities, practices, procedures, processes, and resources for developing, implementing, achieving, and maintaining the environmental policy" (ISO 1996 cited in Federal Facilities Council Report 1999).

ISO 14001 Standard

ISO 14001 sets out the criteria for an Environmental Management System (EMS). It does not state requirements for environmental performance, but maps out a framework that a company or organization can follow to set up an effective EMS. It can be used by any organization that wants to improve resource efficiency, reduce waste, and drive down costs. Using ISO 14001 can provide assurance to company management and employees as well as external stakeholders that environmental impact is being measured and improved. ISO 14001 can also be integrated with other management functions and assists companies in meeting their environmental and economic goals.

ISO 14001, as with other ISO 14000 standards, is voluntary (IISD 2010), with its main aim to assist companies in continually improving their environmental performance, while complying with any applicable legislation. Organizations are responsible for setting their own targets and performance measures, with the standard serving to assist them in meeting objectives and goals and in the subsequent monitoring and measurement of these (IISD 2010).

The standard can be applied to a variety of levels in the business, from organizational level, right down to the product and service level (RMIT university). Rather than focusing on exact measures and goals of environmental performance, the standard highlights what an organization needs to do to meet these goals (IISD 2010).

ISO 14001 is known as a generic management system standard, meaning that it is relevant to any organization seeking to improve and manage resources more effectively. This includes:

- single-site to large multi-national companies

- high-risk companies to low-risk service organizations

- manufacturing, process, and the service industries, including local governments

- all industry sectors including public and private sectors

- original equipment manufacturers and their suppliers.

All standards are periodically reviewed by ISO to ensure they still meet market requirements. The current version ISO 14001:2004 was last reviewed in 2012. The ISO committee decided a revision was necessary. The new version is expected by the end of 2015. After the revision has been published, certified organizations get a three-year transition period to adapt their environmental management system to the new edition of the standard. The new version of ISO 14001 is going to focus on the improvement of environmental performance rather than to improve the management system itself.

The new ISO 14001:2015 standard has been published and includes several new updates all aimed at making environmental management more comprehensive and relevant to the supply chain. One of the main updates is to consider environmental impacts during the life cycle — although there is no requirement to actually complete a life cycle analysis. In addition the commitments of top management and the methods of evaluating compliance have also been strengthened.

Another significant change has been the linking of ISO 14001 to the general management system structure, introduced in 2015, called the High Level Structure. Both ISO 9001 and 14001 use this same structure making implementation and auditing more uniform. The new standard also requires the holder of the certificate to specify risks and opportunities and how to address them.

Plan: establish Objectives and Processes Required

Prior to implementing ISO 14001, an initial review or gap analysis of the organization's processes and products is recommended, to assist in identifying all elements of the current operation and, if possible, future operations, that may interact with the environment, termed "environmental aspects" (Martin 1998). Environmental aspects can include both direct, such as those used during manufacturing, and indirect, such as raw materials (Martin 1998). This review assists the organization in establishing their environmental objectives, goals, and targets, which should ideally be measurable; helps with the development of control and management procedures and processes; and serves to highlight any relevant legal requirement, which can then be built into the policy (Standards Australia/Standards New Zealand 2004).

Do: implement the Processes

During this stage, the organization identifies the resources required and works out those members of the organization responsible for the EMS' implementation and control (Martin 1998). This includes establishing procedures and processes, although only one documented procedure is specified related to operational control. Other procedures are required to foster better management control over elements such as documentation control, emergency preparedness and response, and the education of employees, to ensure that they can competently implement the necessary processes and record results (Standards Australia/Standards New Zealand 2004). Communication and participation across all levels of the organization, especially top management, is a vital part of the implementation phase, with the effectiveness of the EMS being dependent on active involvement from all employees.

Check: measure and Monitor the Processes and Report Results

During the "check" stage, performance is monitored and periodically measured to ensure that the organization's environmental targets and objectives are being met (Martin 1998). In addition, internal audits are conducted at planned intervals to ascertain whether the EMS meets the user's expectations and whether the processes and procedures are being adequately maintained and monitored (Standards Australia/Standards New Zealand 2004).

Act: take Action to Improve Performance of EMS Based on Results

After the checking stage, a management review is conducted to ensure that the objectives of the EMS are being met, the extent to which they are being met, and that communications are being appropriately managed; and to evaluate changing circumstances, such as legal requirements, in order to make recommendations for further improvement of the system (Standards Australia/ Standards New Zealand 2004). These recommendations are incorporated through continual improvement: plans are renewed or new plans are made, and the EMS moves forward.

Continual Improvement Process (CI)

ISO 14001 encourages a company to continually improve its environmental performance. Apart from the obvious – the reduction in actual and possible negative environmental impacts – this is achieved in three ways:

- Expansion: More and more business areas get covered by the implemented EMS.

- Enrichment: More and more activities, products, processes, emissions, resources, etc. get managed by the implemented EMS.

- Upgrading: An improvement of the structural and organizational framework of the EMS, as well as an accumulation of know-how in dealing with business-environmental issues.

Overall, the CI concept expects the organization to gradually move away from merely operational environmental measures towards a strategic approach on how to deal with environmental challenges.

Benefits

ISO 14001 was developed primarily to assist companies with a framework for better management control that can result in reducing their environmental impacts. In addition to improvements in performance, organizations can reap a number of economic benefits including higher conformance with legislative and regulatory requirements (Sheldon 1997) by adopting the ISO standard. By minimizing the risk of regulatory and environmental liability fines and improving an organization's efficiency (Delmas 2009), benefits can include a reduction in waste, consumption of resources, and operating costs. Secondly, as an internationally recognized standard, businesses operating in multiple locations across the globe can leverage their conformance to ISO 14001, eliminating the need for multiple registrations or certifications (Hutchens 2010). Thirdly, there has been a push in the last decade by consumers for companies to adopt better internal controls, making the incorporation of ISO 14001 a smart approach for the long-term viability of businesses. This can provide them with a competitive advantage against companies that do not adopt the standard (Potoki & Prakash, 2005). This in turn can have a positive impact on a company's asset value (Van der Deldt, 1997). It can lead to improved public perceptions of the business, placing them in a better position to operate in the international marketplace (Potoki & Prakash 1997; Sheldon 1997). The use of ISO 14001 can demonstrate an innovative and forward-thinking approach to customers and prospective employees. It can increase a business's access to new customers and business partners. In some markets it can potentially reduce public liability insurance costs. It can serve to reduce trade barriers between registered businesses (Van der Deldt, 1997). There is growing interest in including certification to ISO 14001 in tenders for public-private partnerships for infrastructure renewal. Evidence of value in terms of environmental quality and benefit to the taxpayer has been shown in highway projects in Canada.

Conformity Assessment

ISO 14001 can be used in whole or in part to help an organization (for-profit or not-for-profit) better manage its relationship with the environment. If all the elements of ISO 14001 are incorporated into the management process, the organization may opt to prove that it has achieved full alignment or conformity with the international standard, ISO 14001, by using one of four recognized options. These are:

1. make a self-determination and self-declaration, or

2. seek confirmation of its conformance by parties having an interest in the organization, such as customers, or

3. seek confirmation of its self-declaration by a party external to the organization, or

4. seek certification/registration of its EMS by an external organization.

ISO does not control conformity assessment; its mandate is to develop and maintain standards. ISO has a neutral policy on conformity assessment. One option is not better than the next. Each option serves different market needs. The adopting organization decides which option is best for them, in conjunction with their market needs.

Option 1 is sometimes incorrectly referred to as "self-certify" or "self-certification". This is not an acceptable reference under ISO terms and definitions, for it can lead to confusion in the market. The user is responsible for making their own determination. Option 2 is often referred to as a customer or 2nd-party audit, which is an acceptable market term. Option 3 is an independent third-party process by an organization that is based on an engagement activity and delivered by specially trained practitioners. This option was based on an accounting procedure branded as the EnviroReady Report, which was created to help small- and medium-sized organizations. Its development was originally based on the Canadian Handbook for Accountants; it is now based on an international accounting standard. The fourth option, certification, is another independent third-party process, which has been widely implemented by all types of organizations. Certification is also known in some countries as registration. Service providers of certification or registration are accredited by national accreditation services such as UKAS in the UK.

ISO 14001 and EMAS

In 2010, the latest EMAS Regulation (EMAS III) entered into force; the scheme is now globally applicable, and includes key performance indicators and a range of further improvements. Currently, more than 4,500 organisations and approximately 7,800 sites are EMAS registered.

Complementarities and Differences

ISO 14001's environmental management system requirements are very similar to those of EMAS. Additional requirements for EMAS include:

- stricter requirements on the measurement and evaluation of environmental performance against objectives and targets.

- government supervision of the environmental verifiers

- strong employee involvement; EMAS organisations acknowledge that active employee involvement is a driving force and a prerequisite for continuous and successful environmental improvements.

- environmental core indicators creating multi-annual comparability within and between organisations

- mandatory provision of information to the general public

- registration by a public authority.

ISO 14001 use in Supply Chains

There are many reasons that ISO 14001 should be potentially attractive to supply chain managers, including the use of the voluntary standard to guide the development of integrated systems, its requirement for supply chain members in industries such as automotive and aerospace, the potential of pollution prevention leading to reduced costs of production and higher profits, its alignment with the growing importance of corporate social responsibility, and the possibility that an ISO-registered system may provide firms with a unique environmental resource, capabilities, and benefits that lead to competitive advantage.

Emerging areas of research are starting to address the use of this standard to show that ISO 14001 registration can be leveraged across the supply chain for competitive advantage. By looking at ISO 14001 registered firms, information from the study compared different amounts of integration and sustainability in the supply chain. Several research propositions and an empirical framework posit the impacts of ISO 14001 on supply chain design.

The propositions include:

1. ISO registration leading to more proactive environmental management including process and performance measurement related to sustainability across a supply chain;

2. That ISO-registered plants with formal environmental management systems will have higher levels of communication required between OEMs and Tier I suppliers;

3. ISO-registered plants with direct relationships to other registered plants in their supply chain will have higher levels of waste reduction and cost efficiency than nonregistered plants;

4. ISO-registered plants with direct relationships to other registered plants in the supply chain will have sustainable practices and projects with better ROI than nonregistered firms;

5. ISO-registered plants with direct relationships to other registered plants will have higher levels of customer relationship management and will be positively associated with greater expansion opportunities and image than non-registered plants;

6. ISO-registered plants with direct relationships to other registered plants will have fewer issues with employee health and reduced numbers of safety incidents than nonregistered plants;

7. ISO-registered plants with a direct relationship to other registered plants will have a strong positive relationship between formal communication, training, monitoring/control systems, and firm performance; and

8. ISO-registered plants with a direct relationship to other registered plants will have higher levels of involvement and communication, which will be positively related to more internal and external integration with supply chain members.

List of ISO 14000 Series Standards

- ISO 14001 Environmental management systems - Requirements with guidance for use

- ISO 14004 Environmental management systems - General guidelines on implementation

- ISO 14006 Environmental management systems - Guidelines for incorporating ecodesign

- ISO 14015 Environmental assessment of sites and organizations

- ISO 14020 series (14020 to 14025) Environmental labels and declarations

- ISO 14030 discusses post-production environmental assessment

- ISO 14031 Environmental performance evaluation—Guidelines

- ISO 14040 series (14040 to 14049), Life Cycle Assessment, LCA, discusses pre-production planning and environment goal setting.

- ISO 14046 sets guidelines and requirements for water footprint assessments of products, processes, and organizations. Includes only air and soil emissions that impact water quality in the assessment

A water footprint assessment can assist in:

- assessing the magnitude of potential environmental impacts related to water;

- identifying opportunities to reduce water related potential environmental impacts associated with products at various stages in their life cycle as well as processes and organizations;

- strategic risk management related to water;

- facilitating water efficiency and optimization of water management at product, process and organizational levels;

- informing decision-makers in industry, government or non-governmental organizations of their potential environmental impacts related to water (e.g. for the purpose of strategic planning, priority setting, product or process design or redesign, decisions about investment of resources);

- providing consistent and reliable information, based on scientific evidence for reporting water footprint results.

A water footprint assessment alone is insufficient to be used to describe the overall potential environmental impacts of products, processes or organizations. The water footprint assessment according to this International Standard can be conducted and reported as a stand-alone assessment, where only impacts related to water are assessed, or as part of a life cycle assessment, where consideration is given to a comprehensive set of environmental impacts and not only impacts related to water. In this International Standard, the term "water footprint" is only used when it is the result of an impact assessment. The specific scope of the water footprint assessment is defined by the users of this International Standard in accordance with its requirements.

- ISO 14046 2014, Environmental Management- Water Footprint- Principles, Requirements, and Guidelines

- ISO 14050 terms and definitions

- ISO 14062 Integrating environmental aspects into product design and development (2002)

- ISO 14063 environmental communication guidelines and examples (2006)

- ISO 14064 measuring, quantifying, and reducing greenhouse gas emissions

- ISO 19011 specifies one audit protocol for both 14000 and 9000 series standards together

ISO 14064

The ISO 14064 standard (published in 2006) is part of the ISO 14000 series of International Standards for environmental management. The ISO 14064 standard provides governments, businesses, regions and other organisations with a complimentary set of tools for programs to quantify, monitor, report and verify greenhouse gas emissions. The ISO 14064 standard supports organisations to participate in both regulated and voluntary programs such as emissions trading schemes and public reporting using a globally recognised standard.

Structure of Standard

The Standard is published in three parts:

- ISO 14064-1:2006 specifies principles and requirements at the organization level for quantification and reporting of greenhouse gas (GHG) emissions and removals. It includes requirements for the design, development, management, reporting and verification of an organization's GHG inventory.

- ISO 14064-2:2006 specifies principles and requirements and provides guidance at the project level for quantification, monitoring and reporting of activities intended to cause greenhouse gas (GHG) emission reductions or removal enhancements. It includes requirements for planning a GHG project, identifying and selecting GHG sources, sinks and reservoirs relevant to the project and baseline scenario, monitoring, quantifying, documenting and reporting GHG project performance and managing data quality.

- ISO 14064-3:2006 specifies principles and requirements and provides guidance for those conducting or managing the validation and/or verification of greenhouse gas (GHG) assertions. It can be applied to organizational or GHG project quantification, including GHG quantification, monitoring and reporting carried out in accordance with ISO 14064-1 or ISO 14064-2.

ISO 14064-3:2006 specifies requirements for selecting GHG validators/verifiers, establishing the level of assurance, objectives, criteria and scope, determining the validation/verification approach, assessing GHG data, information, information systems and controls, evaluating GHG assertions and preparing validation/verification statements.

The principles behind ISO 14064 have been used in national calculation methodologies such as the UK's Carbon Trust Standard.

Ems Standard: Iso 14000 Series

Companies throughout the world have begun to understand the requirements of an environmental management system (EMS) as well as the benefits such a system can provide. However, given the nature of company dynamics, changes to business priorities, personnel and work patterns, it is imperative to introduce a proper management system which maintains the focus on environmental improvement through the changing landscape of corporate activity. It is against this background that ISO 14000 Series are being developed and implemented.

Evolution

The International Organisation for Standardisation (ISO) was formed in 1947 and has since become the premier international standards organisation. Since its inception, ISO's mission has been to promote worldwide standardisation in order to facilitate international commerce. The organisation began by developing international agreements and in 1951 published its first standard, Standard Reference Temperature for Industrial Length Measurement. Since that time, ISO has developed more than 9,000 standards for a variety of subjects ranging from screw threads and fasteners to high-tech clean rooms. ISO also developed the widely used quality management system (QMS) standards, i.e., the ISO 9000 series. The ISO standards are published for voluntary acceptance, but they are often incorporated into national standards of individual countries.

The membership of ISO includes over 100 countries. Each Member Country is represented by one standards organisation. For example, the American National Standards Institute (ANSI) represent the U.S and Great Britain by the British Standards Institute (BSI).

ISO considers the following three key principles in developing international standards:

(i) Consensus: The views of all interested parties are taken into account, including manufacturers, vendors, consumer groups, testing laboratories, governments, engineering professionals and research organisations.

(ii) Industry-wide applicability: The goal is to draft standards that satisfy industries and customers worldwide (ISO has no authority to impose its standards on any government or organisation).

(iii)Voluntary nature: All of the international standards developed are voluntary. Thus, their acceptance by industry is market-driven and based on voluntary involvement of all interests in the marketplace.

A review of historical developments reveals that standardisation on a worldwide basis was accelerated with quality and occurred generally independent of environmental management. Attempts to standardise quality requirements were made by many organisations. However, it wasn't until the 1990s that a significant level of agreement was reached.

In 1990, Business Charter for Sustainable Development (BCSD), an organisation of fifty business leaders with interest in environment and development issues, was created for environment protection with the premise that economic development can take place only in a healthy environment. Partly in response to the proliferation of differing environmental standards, such as EMAS worldwide, the ISO formed a Strategic Action Group on the Environment (SAGE) in 1991.The purpose of SAGE was to investigate the usefulness of drafting international standards for environmental management. SAGE focused its attention on the following three areas:

(i) Promoting a common worldwide approach to environmental management in business and industry.

(ii) Increasing the ability of incentives for organisations to measure and attain improvements in environmental performance.

(iii)Facilitating world trade and removing potential environmental trade barriers.

As a result of the findings of SAGE, the ISO formed Technical Committee 207 (TC 207) in 1993. TC 207 became responsible for drafting the ISO 14000 series of standards. At early meetings of TC 207, more than thirty countries and 200 representatives expressed a desire to develop new EMS standards. TC 207 itself had members representing some sixty-nine countries and was divided into six sub-committees. These members included representatives from various industries, standards organisations, governments, environmental organisations, and other interest groups.

For some, the motivation for the development of these standards was due to the fear that the increased number of inconsistent national and regional EMS standards would create trade barriers. There was also the concern that the EMAS programme already in practice in Europe would influence the ISO 14000 standards to make them comparable to EMAS. This concern was in part fueled by the fact that the ISO 14000 standards were being developed in conjunction with CEN, i.e., the European standards-setting body. In fact, the ISO set the standards drafting timetable at 30 months, in part because CEN had agreed to accept ISO 14000, if they were finalised quickly enough. There were also fears that the standards were a reflection of European not American technology. Nevertheless, the standards were developed with each participating member having equal say in the process.

A sub-committee of TC 207 prepared the draft EMS Standard, based on BS 7750. It was presented at the Earth Summit in Rio de Janeiro, and the draft of the standard was agreed in Oslo (June 1995).

Principles and Structure

The aim of the ISO 14000 series of standards is to help organisations implement and improve their EMS.

Some of the principles governing ISO 14000 series are (Tech Monitor, Vol. 12. No. 5 Sept-Oct 1995):

(i) Understand all activities and processes being undertaken by the organisation.

(ii) Identify potential aspects associated with the activities of the unit at all stages and determine their impact on the environment.

(iii) Determine processes/procedures/operation steps that can be controlled to eliminate or minimise the likelihood of an occurrence of the adverse impact.

(iv) Identify the regulatory requirements relating to them and establish target level and tolerances, which must be met to ensure that operations affecting the environment are under control.

(v) Establish a monitoring mechanism to ensure control of these aspects.

(vi) Establish corrective actions to be taken when monitoring indicates that a particular aspect is not under control.

(vii) Establish a system of emergency preparation and for meeting such exigencies.

(viii) Establish procedures for verification to confirm that the environmental management system is working effectively in compliance with regulations and recording continuous improvement.

(ix) Establish documentation concerning all procedures and records appropriate to these principles and their application.

Given the importance of the stakes involved and the generic nature of the international requirements covered, high expectations have been placed on the ISO 14000 series standards. The standards can be classified into the following two categories based on their focus:

(i) Organisation or process standards: These include environmental management system (EMS), environmental auditing (EA) and environmental performance evaluation (EPE).

(ii) Product-oriented standards: These include life-cycle assessment (LCA), environmental labelling (EL) and environmental aspects in product standards (EAPS).

The ISO 14000 series standards are of two types, and these are: (i) Normative standards: These indicate the requirements that must be met and can be audited for certification. (ii) Informative standards: These provide guidance and the requirements need not be audited for certification.

In the 14000 series of standards, ISO 14001, (i.e., Environmental Management Systems-specification with guidance for use) is the only normative standard and all other standards are informative standards intended to support the implementation of EMS.

As regards the structure of the Series, ISO 14001, i.e., the requirements for EMS, forms the nucleus, and in the first orbit is ISO 14004, i.e., the guideline standard. The other standards in the series are supporting systems.

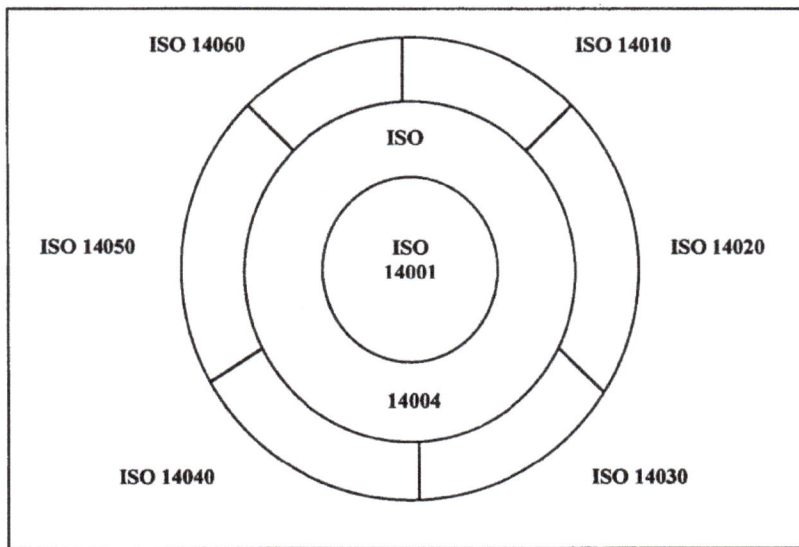

Structure of ISO Series and Inter-relationships

Supporting Systems

The supporting systems of ISO 14000 consist of ISO 14010, ISO 14020, ISO 14030, ISO 14040, ISO 14050 and ISO 14060.

Consider the table, which gives a list of 150 14000 Series standards with their respective publication dates.

ISO 14000 Series Standards

ISO No.	Title of International Standard/Guideline/Technical Report
ISO 14001	Environmental management
ISO No.	Title of International Standard/Guideline/Technical Report
	systems – Specification with guidance for use.
ISO 14004	Environmental management systems – General guidelines on principles, systems and supporting techniques.
ISO/AWI 14004	Revision of ISO 14004:1996.
ISO 14010	Guidelines for environmental auditing – General principles.
ISO 14011	Guidelines for environmental auditing - Audit procedures – Auditing of environmental management systems.

ISO 14012	Guidelines for environmental auditing – Qualification criteria for environmental auditors.
ISO 14015	Environmental management – Environmental assessment of sites and organisations (EASO).
ISO/DIS 19011	Guidelines for quality and/or environmental management systems auditing.
ISO 14020 2nd Edition	Environmental labels and declarations – General principles.
ISO 14020: 1998/DAM1	Draft amendment 1 to ISO 14020:1998.
ISO 14021	Environmental labels and declarations - Self-declared environmental claims (Type II environmental labelling).
ISO 14024	Environmental labels and declarations - Type I environmental labelling – Principles and procedures.
ISO/TR 14025	Environmental labels and declarations – Type III environmental declarations.
ISO 14031	Environmental management – Environmental performance evaluation – Guidelines.
ISO/TR 14032	Environmental management – Examples of environmental performance evaluation.
ISO No.	Title of International Standard/Guideline/Technical Report
ISO 14040	Environmental management – Life cycle assessment – Principles and framework.
ISO 14041	Environmental management – Life cycle assessment – Goal and scope definition and inventory analysis.
ISO 14042	Environmental management – Life cycle assessment – Life cycle impact assessment.
ISO 14043	Environmental management – Life cycle assessment – Life cycle interpretation.
ISO/WD TR 14047	Environmental management – Life cycle assessment – Examples of application of ISO 14042.
ISO/CD TR 14048	Environmental management – Life cycle assessment – Life cycle assessment data documentation format.

ISO/TR 14049	Environmental management – Life cycle assessment – Examples of application of ISO 14041 to goal and scope definition and inventory analysis.
ISO 14050	Environmental management – Vocabulary.
ISO 14050: 1998/DAM1	Draft amendment 1 to ISO 14050:1998.
ISO/TR 14061	Information to assist forestry organisations in the use of Environmental Management System standards ISO 14001 and ISO 14004.
ISO/AWI 14062	Guidelines for integrating environmental aspects into product development.
ISO Guide 64	Guide for the inclusion of environmental aspects in product standards.

Environmental Auditing (ISO 14010, ISO 14011 and ISO 14012)

An environmental audit is defined as the systematic documented verification process of objectively obtaining and evaluating audit evidence to determine whether specified environmental activities, processes, conditions, management systems, or information about these matters conform with audit criteria, and communicating the results of this process to the clients. The guiding principles of environmental audits include:

- basing the audit on defined objectives and drawing inferences based on analysis, interpretation and documentation of appropriate information;

- utilising an audit team that is independent of the activities they audit and utilising an auditor who meets the specifics of qualification criteria;

- exercising of due professional care by the auditor to maintain confidentiality and adequate quality assurance;

- using appropriate procedures for an objective audit;

- developing audit criteria, evidence and findings;

- ensuring that the process provides a desired level of confidence in the reliability of the audit findings and conclusions;

- providing an adequate report of findings.

Note that while ISO 14010 provides guidance on general principles for conducting environmental audits, ISO 14011 elaborates on the framework for the conduct of audit in order to ascertain if the organisation is doing what it says it will do and whether the EMS conforms to ISO 14001. ISO 14012 provides guidance on auditor qualification criteria, including education, training, work experience, personal attributes and skills, maintenance of competence and due diligence.

Environmental Labelling Standards (ISO 14020)

ISO 14020 is based on a voluntary environmental labelling standard that requires a third party verification and is designed to reduce burdens arising from diverse multiple labelling. Existing eco-labelling schemes, based on government initiatives, aim at influencing customer decisions to select environmentally friendly products, but they lack the application of uniform criteria. These are, therefore, difficult to comply with and are potential trade barriers. The coverage of ISO 14020 is broad and includes goods or services for consumer, commercial and industrial purposes.

The guiding principles and practices for third-party environmental labelling programmes include the following:

- Standard and criteria applicable for environmental labels must be developed through a consensus process, and the programme must be voluntary.

- A product can be considered for environmental regulations of the country in which it is manufactured and marketed.

- Environmental labelling programmers should be selective and should distinguish leading product alternatives.

- The product criteria developed by the certification agency should be periodically reviewed to account for new developments and technologies.

- The process should be transparent with regard to criteria, certification and award procedures.

- The criteria should be revised periodically and clearly demonstrate that the funding sources for the programme do not create conflicts of interest or undue influence.

- The labelling programme should use scientific and reproducible methodologies to assess the environmental impacts of products.

- Labelling programmes should be accessible, objective and affordable.

The requirements for awarding a label are divided into general rules that apply to all products and applications, and specific product criteria that set requirements for each product category. The specific product criteria are the only criteria that may be considered as a basis for awarding the label. The certification agency awards the label when satisfied that the applicant has complied with the specific product criteria for the category. It maintains a publicly available list of products currently licensed to carry the label.

After the label has been awarded, it is the responsibility of the certification agency to take all necessary steps to ensure ongoing compliance with the product criteria. The certification agency will require the licensee to take corrective action, if monitoring indicates that compliance is not being maintained.

Environmental Performance Standards (ISO 14030)

The environmental performance evaluation (EPE) is an important ongoing internal management process. It uses environmental indicators to compare an organisation's past and present environmental performance with it's environmental objectives, targets or other intended levels of environmental performance. This process helps management to measure, analyse, assess, report and communicate an organisation's environmental performance over time and to determine necessary actions.

The environmental performance evaluation standard is based on Edward Deming's PDCA cycle, i.e., Plan (P), Do (D), Check (C) and Act (A). Planning a EPE involves considering management priorities and selecting environmental indicators before evaluating environmental performance (D) and reviewing and improving EPE (C). The EPE process involves actions taken (A) to improve the systems, operations, processes and environmental performance.

EPE standards provide guidelines on identification and selection of environmental indicators. There are two types of environmental indicators, and these are:

(i) Environmental performance indicators (EPI): These include the following:

- the people, practices and producers at all levels of the organisation;

- the design, operation and maintenance of, supply to, and delivery from, an organisation's facilities and equipment;

- the materials, energy, products, services, waste and emissions related to the organisation's operations and activities.

(ii) Environmental condition indicators (ECI): These are used to describe the conditions of the environment in relation to the organisation.

Life Cycle Assessment (ISO 14040)

ISO 14040 provides guidelines for incorporating life cycle assessment (LCA) into environmental management programmes (recall that in Unit 6 we discussed LCA in detail). An LCA is defined as a compilation and evaluation of the inputs, outputs and the potential environmental impacts of a product system throughout its life cycle. It is, thus, a systematic set of procedures for compiling and examining the inputs and outputs of materials and energy and the associated environmental impacts directly attributable to the functioning of a product or service system throughout its life cycle from the acquisition of raw materials through final disposal. An LCA is done in order to get the whole picture of the environmental impacts throughout the lifetime of products and services. In other words, an LCA provides significantly more information than does evaluating the impact from the manufacturing process alone. It also provides a systematic way to evaluate the costs and benefits associated with products or service changes at various points in the life cycle.

ISO 14040 covers the following three phases of LCA:

- establishing the goals and the scope of the assessment;

- conducting inventory analysis.

- conducting impact and improvement assessment.

The goals of the LCA study should include the reasons for carrying out the study, the intended applications, the intended audience, the initial data, quality objectives and the type of critical review that will be conducted for the LCA. The scope should include background information for the product or service being evaluated, boundaries of the study, method of impact assessment, data requirements, assumptions and limitations of the study.

Environmental Terms and Definitions (ISO 14050)

The successful operation of the environmental management system requires a correct understanding of the meaning of terms used by all stakeholders in a similar manner. Put differently, communication is important in implementation and in the operation of the environmental management system. This communication will be most effective, if there is a common understanding of the terms used. Many environmental terms and definitions are the result of a recently developed process. The gradual evolution of these environmental concepts invariably means that environmental terminology will continue to develop.

EMS Specification Standards: ISO 14001

This standard is the core of the ISO 14000 series. ISO 14001 presents the elements of a EMS that organisations are required to conform with, if they seek third-party certification. As mentioned previously, ISO 14001 is the only standard in the ISO 14000 series intended to be used for third-party certification. All the other standards in the series are intended as guidance only. In development of your EMS, ISO 14001 will be your most important document – you will refer to it frequently.

The standard consists of the EMS specification in 17 general requirements (referred to as clauses) and three annexes.

Clauses

Because the standard has been written to be applicable to all types of organisations and to accommodate diverse geographical, cultural and social conditions, the requirements or clauses, are generalised and are neither specific nor prescriptive. That is to say, they are descriptive outlining the desired outcome of the system, but do not prescribe specific approaches that an organisation must implement to get there.

Annexes

Annex A provides additional information on the requirements of ISO 14001 and is intended to help avoid any misinterpretation of the actual specification. It is expected that in future revisions to ISO 14001, the information in Annex A may be either incorporated into the clauses or moved into ISO 14004. Annex B identifies links and broad technical correspondences between ISO 14001 and ISO 9001 in a tabular form. The objective of this annex is to demonstrate how a EMS could be combined with a quality management system. Annex C simply presents a bibliography of documents from both the ISO 9000 and ISO 14000 series.

Note that these annexes are solely for information and not for third-party certification. ISO 14001 is a management system standard. It is not a performance or product standard, although the framers developed it with the idea that the result of its implementation should give better industry environmental performance. The standard represents a shift towards holistic proactive management and total employee involvement. ISO 14001 urges employees to define their environmental roles from the bottom up and requires the support of senior management. It is a comprehensive framework that contains core elements for managing an organisation's processes and activities to identify significant environmental aspects the organisation can control, and over which it can be expected to have an influence. Any organisation or facility or organisation of any size anywhere in the world can use the standard.

An EMS conforming to ISO 14001 contains the following elements:

- An environmental policy supported by senior management.

- Identification of environmental aspects and significant impacts.

- Identification of legal and other requirements.

- Environmental goals, objectives and targets that support the policy.

- An environmental management programme.

- Definition of roles, responsibilities and authorities.

- Training and awareness procedures.

- Process for communication of the EMS to all interested parties.

- Document and operational control procedures.

- Procedures for emergency response.

- Procedures for monitoring and measuring operations that can have a significant impact on the environment.

- Procedures to correct non-conformance.

- Record management procedures.

- A programme for auditing and corrective action.

- Procedures for management review.

Note that while ISO 14001 is more likely to be used in manufacturing or processing industries, it can also be applied to services such as construction, architecture, health care and engineering. This standard essentially requires an organisation to state what it does in environmental management and do what it states. ISO 14001 is neither a product standard nor an environmental performance standard. It does not require an organisation to establish or disclose performance or audit results and does not require certification.

Let us now discuss the guidelines for planning and implementation of ISO 14001.

Implementation oF Ems Conforming To Iso 14001

The key to a successful ISO 14001 EMS is having documented procedures that are implemented and maintained in such a way that successful achievement of environmental goals is commensurate with the nature and scale of the activities.

There are three components to a EMS and these are:

(i) A written programme: The written programme requires an organisation to be committed to producing a quality product with the lowest possible environmental impact and sets forth the procedures to be followed to achieve this goal.

(ii) Education and training: The implementation will not be effective, unless all employees have access to and understand the EMS, and this is achieved through education and training.

(iii) Knowledge of relevant local and central environmental regulations: The EMS must incorporate the applicable environmental legal requirements.

In addition, the EMS must include appropriate monitoring and review to ensure effective functioning of the EMS and to identify and implement corrective measures in a timely manner.

The various steps involved in the establishment of a EMS are as follows:

(i) Obtain management commitment: Senior management can demonstrate its commitment by playing an active, visible role in the EMS implementation process, providing funding and allocating resources, and promoting employee awareness and motivation. A EMS should be viewed as a tool to achieve continuous environmental improvement, rather than daily fire-fighting just to meet the mandatory requirements.

(ii) Review current environmental programmes: Once the management commitment is secured, the next step is to conduct and document a preliminary review of the organisation's current environmental programmes and management systems. The process includes reviewing existing environmental management policies, operating procedures and training programmes as well as the methods for identifying regulatory requirements.

(iii) Conduct gap analysis: Organisations need to evaluate as to what extent the existing procedures conform to the requirements of ISO 14001. Such a gap analysis identifies the actions necessary to build a EMS. To do this, organisations would need to compare their existing procedures to the following main stages of ISO 14001:

- Environmental policy.

- Planning.

- Implementation and operation.

- Checking and corrective action.

- Management review.

- Continuous improvement.

(iv) Identify environmental aspects: All parts of the planning phase contribute to identifying the significant environmental aspects of the operations. An environmental aspect refers to any element of an organisation's activities, products and services that can interact with the environment. ISO 14001 also refers to impacts. Impacts are the actual or potential changes to the environment resulting from any of the environmental aspects. Thus, the relationship between aspects and impacts is one of cause and effect. One of the best ways to identify the environmental aspects of the operation is by developing a process map or flowchart. This exercise involves mapping every step of the process and the inputs and outputs associated with each step. Developing a process map is best accomplished as a team effort, and therefore, representatives of a variety of departments and at diverse levels should be included to produce the most accurate description of the inputs and outputs. Note that an organisation's EMS is a team effort.

(v) Identify significant aspects: An organisation can utilise the expertise of staff to determine the significant environmental aspects of its operations. This will help in focusing on those aspects over which the organisation has some control or influence. Determining which aspects are significant is based on the judgement of severity of the impact caused by the aspect and the frequency of its occurrence, and therefore, each aspect needs to be scrutinised. It requires many iterations, before the team arrives at a final list of significant environmental aspects.

(vi) Decide the environmental policy: Having identified the significant environmental aspects and established the means and norms to improve environmental performance in the form of objectives and targets, organisations need to write their broad intentions concerning the environment. This will form or evolve into an appropriate policy, meeting the requirements of the standards. This document must include the organisation's commitments towards complying with legal requirements, pollution prevention and continuous improvement. Note that policy framing may not be a onetime affair. It is quite likely that organisations are required to revisit the objectives formulated and targets identified commensurating with the outcomes of the initial implementation process. This iterative process helps in firming up the policy statements.

(vii) Identify environmental management programmes: An environmental management programme (EMP) may be considered a project or a plan of action to achieve objectives and targets. At this stage, the methodology of realising the objectives, the steps involved, the persons responsible for implementing these steps, the time frame of implementation, the means required, etc., should be noted.

(viii) Identify training needs and personnel whose work affects the environment: It is essential to identify the training needs of the personnel by comparing the present competence (based on the educational qualifications, past work experience, etc.) with the required competence. Organisations are expected to have training needs, training plans, training modules and training records, and must check the effectiveness of the training imparted. The training must make the personnel aware of the importance of conformance with environmental policy, environmental impacts of their work activity and the consequences of departing from the specified operating procedures.

(ix) Establish communication system: Proper communication channels are imperative to impress on the employees, as well as external interested parties, the reasons for adopting EMS.

(x) Create proper documentation: The EMS documentation describes the interaction among the core elements of the system and provides the key information necessary to understand the environmental management systems in place at the organisation. This is of immense use to the new employees and third-party auditors.

(xi) Conduct EMS audits: Organisations are required to conduct EMS audits to verify the EMS implementation. But, EMS audits conducted merely to meet the requirements of the ISO standard will not be much use in the long term. For sustained benefits, organisations must invest in EMS audit programmes by giving employees proper training, providing appropriate time for audit, etc. Importantly, senior management should demonstrate its commitment to audits and provide visible support to the auditors. This will encourage the auditors to point out the weaknesses in the system without any hesitation. As inept audits – as a result of doctored auditing – are not uncommon, the support of senior management is very critical in carrying out and documenting EMS audits.

(xii) Conduct management review: Management must periodically review the EMS to evaluate its suitability and effectiveness.

(xiii) Perform a final gap analysis: Final gap analyses help organisations determine whether they are ready for a registration audit. In other words, the results of the final gap analyses help organisations decide whether or not to apply for certification by outside auditors.

Registration

Organisations are registered by outside auditors – also known as registrars – meeting the ISO 14001 standard. The registration can be for a specific site, several sites or the entire organisation. The registration process involves the following steps:

(i) Establishing the scope of the registration. This can be done internally or with the help of outside auditors.

(ii) Applying for registration. This is a written application to the registration body.

(iii) Reviewing documents about the organisation's EMS and submitting them to the registration body. Outside auditors perform full assessment.

(iv) Getting the grant of registration by the certification body based on the auditor's findings.

(v) Receiving either approval (conditional or provisional) or disapproval.

(vi) Receiving a certificate as a proof of registration and getting listed in a certification register.

(vii) Maintaining the registration through ongoing activities including monitoring and measuring the EMS, investigating and handling non-conformances, implementing corrective and preventive actions, maintaining environmental records and establishing and maintaining the ISO 14000 audit programme.

The time required to implement ISO 14001 EMS depends on the current status. That is to say, if an organisation's environmental programmes are effective and efficient and they are willing to commit resources, the EMS development will be fast. If all the factors required for the EMS development are in place, it is possible to complete the entire process till certification in eight to ten months. (Note that it may take the equivalent of one fourth of an employee's time for preliminary planning and exploring issues. The final four to five months will be required for checking and improving the established system. The longest and most timeconsuming activity in the entire process will be training.)

Before you read any further, you must note that:

- ISO does not assess or audit environmental management systems to confirm that they conform to ISO 14001.

- ISO does not issue ISO 14000 certificates.

- ISO does not carry out ISO 14000 certificates (these are issued by certification (registration) bodies independently of ISO).

- The ISO logo is a registered trademark and ISO does not authorise its use in connection with ISO 14000 certification (registration).

- ISO 14001 is not a label signifying a green or environmentally friendly product. (No product label, advertisement or other promotional material should give the impression that a product is "ISO 14001-certified" or "ISO 14001-registered". Advertisements, which imply a quality ranking, i.e., ISO 9001, ISO 9002 and ISO 9003, are misleading.)

Benefits of Implementing Iso 14001: An Indian Scenario

Confederation of Indian Industries (CII) is the nodal agency for ISO 14000 in India. By 2001, over 350 organisations in India have established EMS conforming to ISO 14001.

Sector-Wise Distribution Of 14001 Certified Enterprises:

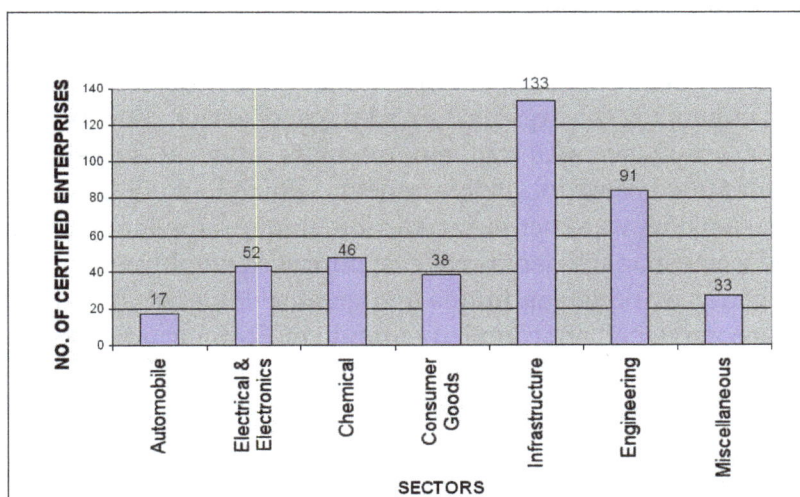

Based on the reports of CII, we will list below the benefits perceived by the organisations after implementing ISO 14001 environmental management system:

- Environmental benefits: These include identification of new opportunities for environmental improvements and improvement in resource conservation and waste minimisation.

- Cost containment and cost savings: These include identification of new opportunities for cost savings and areas of improvement.

- Regulatory compliance: This includes risk reduction and a lower rate of regulatory enforcement.

- Competitiveness: This includes an increase in market share and a decrease in cost of production.

- International trade: This includes an increase in exportrelated business opportunities.

- Greening the suppliers: These include the opportunities for greening the supplier chain. (Note that this has been a major thrust area for those firms who have a wide range of suppliers. At present, most firms generally do not include environmental issues in their purchase contracts. Firms have to impart education/training and transfer of know-how, before actually asking the supply chain to go green. Before long, the suppliers' environmental performance will become important for maintenance of quality and compliance of environmental standards.)

- Other benefits: These include achievement of commitment to good environmental performance at all levels, safe and healthy working conditions as well as better housekeeping.

Some specific cases of firms that have benefited from the establishment of EMS conforming to ISO14001 are given below:

- The case of an international electrical and electronic goods giant. It has a very comprehensive and ambitious EMS programme for more than 200 of its factories worldwide supported by the parent body. Other than ISO 14001, it also conforms to a headquarter-driven quality programme. The environmental aspects recognised by the organisation were development and production of different types of capacitors; manufacture of lamps along with their components and manufacture of audio products, PCBs and electronic and electromagnetic ballast. The most important motivation factors were directions from the headquarters; enhancing the organisation's image; resource conservation benefits and ensuring better compliance with legal requirements. After the implementation of EMS, the organisation identified some intangible benefits, which had not been included in its original goal. These benefits were better interaction among suppliers, employees and authorities; improved waste management practices; improved employee awareness and enhanced staff morale; better working conditions and greater domestic share. The major tasks the organisation accomplished after implementation of EMS included 50% reduction in water consumption; 12 - 25% energy saving depending upon the plant; reduction in packaging mass and reductions in emissions by 95%.

- The case of an organisation was certified to Responsible Care before it was certified to the ISO 14001 protocol. The organisation achieved cost savings from pollution prevention, energy conservation and waste minimisation. The savings included recovery of various products (Rs. 5,55,000), reduction in water, fuel, electricity consumption (Rs. 17,10,000), reuse (Rs. 7,00,000), and others. Note that though the organisation followed the Responsible Care norms, it was better able to recognise potential areas for cost savings through the adoption of ISO 14001.

- The case of a mining organisation that maintains the heavy hydraulic machinery/vehicles used for its excavation work in a workshop situated on top of a hill. As part of the maintenance, oil changes were made and oil was carelessly being thrown on the land, thus contaminating it. Through the EMS, an organisation-wide oil balance study was initiated and it was found that there was a large gap between the oil received and the waste oil generated by the workshop. On a closer study, it was found that the gap accounted for the oil spill on the land. Consequently, the operating practice was changed and trays were installed to collect the waste oil. From the trays, the waste oil was transferred to drums and were sold. This sale resulted in a revenue flow of Rs. 36 lakhs in the first year of the organisation's EMS implementation.

- The case of a paint-manufacturing firm. At this time, there was no monitoring of the amount of water consumed, and different estimates of water consumption ranging between 250 to 400m3 /day were available. Through the EMS, the organisation actually found that it was 490 m3 /day. It decided to make a significant resource saving by reducing water wastage. After plugging the water leakage points and controlling the overflow in the cooling tower, water consumption was reduced to 210 m 3 /day, i.e., a saving of 270m3 /day.

- The case of an electrical cable manufacturing organisation. This organisation makes aluminum core cable with an XLPE (cross-linked polyethylene) coating, which is a black hard plastic material. In the process of making this aluminum core, there was the mechanical drawing of the wire from a metal block. This operation generates fine aluminum dust and was inhaled by the workers in the plant (as this metal dust was not contained), which caused respiratory problems and further led to the absenteeism of the workers. Through the EMS, the organisation decided to contain this waste generation by encasing the metal drawing operation. The aluminum dust generated was properly collected and sold to a local paint manufacturer who used it to manufacture silver paint.

- The case of a leading cloth mill. This organisation is the largest denim-manufacturing firm in India. It was the first denim manufacturer in India to gain the ecologically optmised fabric (EOF) trademark an eco-tex certification. It exports nearly 75 – 80% of its total denim production. The most important motivating factors for the firm were enhancing the organisation's image; augmenting management systems culture from quality to other areas of social concern and ensuring legal compliance and achieving higher environmental performance standards. During the implementation phase, the organisation's main focus was on resource conservation. The benefits that the firm realised after the implementation of ISO 14001 include graduation from ad hoc arrangements to using systems approach for environmental management that is consistent with the overall operations of the unit, incorporation of environmental parameters in the corporate decision-making process and enhancement of the organisation's image and greater worker motivation.

OHSAS 18001

OHSAS 18001, Occupational Health and Safety Assessment Series, (officially BS OHSAS 18001) is an internationally applied British Standard for occupational health and safety management systems. It exists to help all kinds of organizations put in place demonstrably sound occupational health and safety performance. It is a widely recognized and popular occupational health and safety management system.

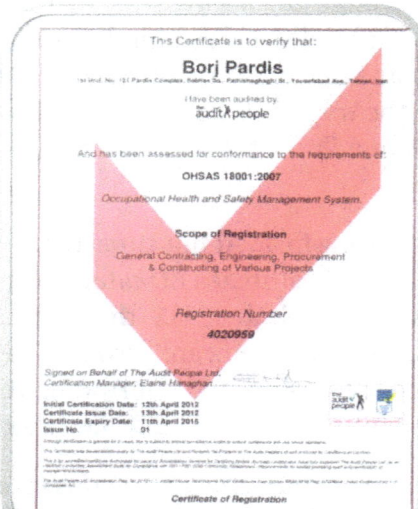

Certificate of conformity to OHSAS 18001:2007

Organizations worldwide recognize the need to control and improve health and safety performance and do so with occupational health and safety management systems (OHSMS). However before 1999 there was an increase of national standards and proprietary certification schemes to choose from. This caused confusion and fragmentation in the market and undermined the credibility of individual schemes.

Development

Recognizing this deficit, an international collaboration called the Occupational Health and Safety Assessment Series (OHSAS) Project Group was formed to create a single unified approach. The Group comprised representatives from national standards bodies, academic bodies, accreditation bodies, certification bodies and occupational safety and health institutions, with the UK's national standards body, BSI Group, providing the secretariat.

Drawing on the best of existing standards and schemes, the OHSAS Project Group published the OHSAS 18000 Series in 1999. The Series consisted of two specifications: 18001 provided requirements for an OHS management system and 18002 gave implementation guidelines. As of 2005, around 16,000 organizations in more than 80 countries were using the OHSAS 18001 specification. By 2009 more than 54,000 certificates had been issued in 116 countries to OHSAS or equivalent OHSMS standards.

Adoption as British Standard

The OHSAS 18001 specification was updated in July 2007. Among other changes, the new specification was more closely aligned with the structures of ISO 9000 and ISO 14000 so that organizations could more easily adopt OHSAS 18001 alongside existing management systems. Additionally, the "health" component of "health and safety" was given greater emphasis.

Later, the BSI Group decided to adopt OHSAS 18001 as a British standard. BSI Group subsequently adopted the updated 18002 guidance specification for publication as BS OHSAS 18002 in 2008.

How They Work

Its supporters claim that an occupational health and safety management system (OHSMS) promotes a safe and healthy working environment by providing a framework that helps organizations to:

- Identify and control health and safety risks

- Reduce the potential for accidents

- Aid legal compliance

- Improve overall performance

The OHSAS 18000 standards provide organizations with the elements of an effective safety management system which can be integrated with other management systems and help organizations achieve better occupational health and safety performance and economic objectives.

BS OHSAS 18001 specifies requirements for an OH&S management system to help an organization develop and implement a policy and objectives, which take into account legal requirements and information about OH&S risks. It applies to all types and sizes of organizations and accommodates diverse geographical, cultural and social conditions.

BS OHSAS 18002 provides guidance for establishing, implementing or improving a management system which is based on OHSAS 18001 and demonstrating successful implementation of OHSAS 18001.

OHSAS 18001 can be aligned with existing ISO 9001 and ISO 14001 management systems. Historically many organizations start with the quality management system ISO 9001, then add the environment management requirements from ISO 14001. Many organizations now look at implementing all three standards at once which can minimize costs and disruption. The standards can be integrated using a standard such as BSI's PAS 99.

Disambiguation

The OHSAS 18000 standards were written and published wholly outside of the International Organization for Standardization (ISO) framework. To avoid confusion, ISO 18000 does exist – but it is a radio-frequency identification standard.

ISO 45001

In October 2013, the International Organization for Standardization approved the project proposal to develop the ISO 45001 standard, an ISO analogue to the OHSAS 18000 standards. This is stated to be taking OHSAS 18001 into account along with other international standards.

OHSAS 18001 and its Comparison with ISO 14001 and ISO 9001

The (OHSAS) specification gives requirements for an occupational health and safety (OH&S) management system, to enable an organisation to control its OH&S risks and improve its performance.

It does not state specific OH&S performance criteria, nor does it give detailed specifications for the design of a management system. OHSAS 18001 has been developed to be compatible with the ISO 9001 (Quality) and ISO 14001 (Environmental) management systems standards, in order to facilitate the integration of quality, environmental and occupational health and safety management systems by organizations.

Scope

This Occupational Health and Safety Assessment Series (OHSAS) specification gives requirements for an occupational health and safety (OH&S) management system, to enable organizations to control its OH&S risks and improve performance. It does not state specific OH&S performance criteria, nor does it give detailed specifications for the design of a management system.

This OHSAS specification is applicable to any organization that wishes to:

a) Establish an OH&S management system to eliminate or minimize risk to employees and other interested parties who may be exposed to OH&S risks associated with its activities;

b) Implement, maintain and continually improve an OH&S management system;

c) Assure itself of its conformance with its stated OH&S policy;

d) Demonstrate such conformance to others;

e) Seek certification/ registration of its OH&S management system by an external organization; or

f) Make a self-determination and declaration of conformance with this OHSAS specification.

This OHSAS specification is intended to address occupational health and safety rather than product and service safety.

Objectives

The organization shall establish and maintain documented occupational health and safety objectives, at each relevant function and level with the organization. This needs to be quantified wherever practicable.

When establishing and reviewing its objectives, an organization shall consider its legal and other requirements, its OH&S hazards and risks, its technological options, its financial, operational, and business requirements, and the views of interest parties. The objectives shall be consistent with the OH&S policy, including the commitment to continual improvement.

A main driver for this was to try to remove confusion in the workplace from the proliferation of certifiable OH&S. specifications. This is based on the concerted effort from a number of the world's leading national standards bodies, certification bodies, specialist consultancies and published documents:

- BS8800:1996 Guide to occupational health and safety management systems

- Technical Report NPR 5001: 1997 Guide to an occupational health and safety management system

- SGS & ISMOL ISA 2000:1997 Requirements for Safety and Health Management Systems

- BVQI SafetyCert: Occupational Safety and Health Management Standard

- DNV Standard for Certification of Occupational Health and Safety Management Systems(OHSMS):1997

- Draft NSAI SR 320 Recommendation for an Occupational Health and Safety (OH and S) Management System

- Draft AS/NZ 4801 Occupational health and safety management systems Specification with guidance for use

- Draft BSI PAS 088 Occupational health and safety management systems

- UNE 81900 series of pre-standards on the Prevention of occupational risks

- Draft LRQA SMS 8800 Health & safety management systems assessment criteria

Application and Benefits

The OHSAS specification is applicable to any organisation that wishes to:

- Establish an OH&S management system to eliminate or minimise risk to employees and other interested parties who may be exposed to OH&S risks associated with its activities

- Implement, maintain and continually improve an OH&S management system

- Assure itself of its conformance with its stated OH&S policy

- Demonstrate such conformance to others

- Seek certification/registration of its OH&S management system by an external organization

- Make a self-determination and declaration of conformance with this OHSAS specification.

Steps to OHSAS 18000 Certification

The process for establishing an OHSAS 18001 management system is essentially the same as the process used of implementing an ISO 14001 system:

- Preliminary review of existing programs and systems

- Identification of hazards and applicable laws and regulations

- Developing new procedures

- Training personnel

- Implementing new programs such as internal audit and management review

- Seeking certification of the OHS program

The OH&S management program(s) needs to be reviewed at regular and planned intervals to address changes to the activities, products, services, or operating conditions of the organization.

Training, Awareness and Competence

The organization shall establish and maintain procedures to ensure that its employees working at each relevant function and level are aware of:

- The importance of conformance to the OH&S policy and procedures, and to the requirements of the OH&S management system;

- The OH&S consequences, actual and potential, of their work activities and the OH&S benefits of improved personal performance;

- Their roles and responsibilities in achieving conformance to the OH&S policy and procedures and to the requirements of the OH&S management system, including emergency preparedness and response requirements

- The potential consequences of departure from specified operating procedures.

Training procedures shall take into account differing levels of: Responsibility, ability and literacy and Risk

Employee involvement and consultation arrangements is to be documented and interested parties informed.

Employees shall be:

- Involved in the development and review of policies and procedures to manage risks;

- Consulted where there are any changes that affect workplace health and safety;

- Represented on health and safety matters; and

- Informed as to who is their employee OH&S representative(s) and specified management appointee.

Documentation

The organization needs to establish and maintain information, in a suitable medium such as paper or electronic form, that:

a) Describes the core elements of the management system and their interaction; and

b) Provides direction to related documentation.

c) Periodically reviewed, revised as necessary and approved for adequacy by authorized personnel;

d) Archival documents and data retained for legal or knowledge preservation purposes or both are suitably identified.

Operational Control

The organization shall identify those operations and activities that are associated with identified risk where control measures need to be applied. The organization shall plan these activities, including maintenance, in order to ensure that they are carried out under specified conditions by:

a) Establishing and maintaining documented procedures to cover situations where their absence could lead to deviations from the OH&S policy and the objectives;

b) Stipulating operating criteria in the procedures;

c) Establishing and maintaining procedures related to the identified OH&S risks of goods, equipment and services purchased and/or used by the organization and communicating relevant procedures and requirements to suppliers and contractors;

d) Establishing and maintaining procedures for the design of workplace, process, installations, machinery, operating procedures and work organization, including their adaptation to human capabilities, in order to eliminate or reduce OH&S risks at their source.

Emergency Preparedness and Response

The organization shall establish and maintain plans and procedures to identify the potential for, and responses to, incidents and emergency situations, and for preventing and mitigating the likely illness and injury that may be associated with them.

The organization shall review its emergency preparedness and response plans and procedures, in particular after the occurrence of incidents or emergency situations. The organization shall also periodically test such procedures where practicable.

Performance Measurement and Monitoring

The organization shall establish and maintain procedures to monitor and measure OH&S performance on a regular basis. These procedures shall provide for:

- Both qualitative and quantitative measures, appropriate to the needs of the organization;

- Monitoring of the extent to which the organization's OH&S objectives are met;

- Proactive measures of performance that monitor compliance with the OH&S management program, operational criteria and applicable legislation and regulatory requirements;

- Reactive measures of performance to monitor accidents, ill health, incidents (including near-misses) and other historical evidence of deficient OH&S performance;

- Recording of data and results of monitoring and measurement sufficient to facilitate subsequent corrective and preventive action analysis

If monitoring equipment is required for performance measurement and monitoring, the organization shall establish and maintain procedures for the calibration and maintenance of such equipment. Records of calibration and maintenance activist and results shall be retained.

OH&S records shall be legible, identifiable and traceable to the activities involved. OH&S records shall be stored and maintained in such a way that they are readily retrievable and protected against damage, deterioration or loss. Their retention times shall be established and recorded.

Audit

The organization shall establish and maintain an audit program and procedures for periodic OH&S management system audits to be carried out, in order to:

a) Determine whether or not the OH&S management system:

i. Conforms to planned arrangements for OH&S management including the requirements of this OHSAS specification;

ii. Has been properly implemented and maintained; and

iii. Is effective in meeting the organization's policy and objectives;

b) Review the results of previous audits;

c) Provide information on the results of audits to management.

The audit program, including any schedule, shall be based on the results of risk assessments of the organization's activities, and the results of previous audits. The audit procedures shall cover the scope, frequency, methodologies and competencies, as well as the responsibilities and requirements for conducting audits and reporting results.

Wherever possible, audits shall be conducted by personnel in dependent of those having direct responsibility for the activity being examined.

The organization's top management shall, at intervals that it determines, review the OH&S management system, to ensure it continuing suitability, adequacy and effectiveness. The management review process shall ensure that the necessary information is collected to allow management to carry out this evaluation. This review shall be documented.

The management review shall address the possible need for changes to policy, objectives and other elements of the OH&S management system, in the light of OH&S management system audit results, changing circumstances and the commitment to continual improvement.

The table analyses the interrelationship among OHSAS 18001, ISO 14001 and ISO 9001

Inter-relationship between OHSAS 18001, ISO 14001 and ISO 9001

OHSAS 18001	ISO 14001	ISO 9001
Scope	Scope	Scope
Reference publications	Normative references	Normative references
Terms and definitions	Definitions	Definitions
OH&S management system elements	Environmental management system requirements	Quality system requirements
General requirements	General Requirements	General (1st sentence)

OH&S Policy	Environmental policy	Quality policy
Planning	Planning	Quality system
Planning for hazard identification, risk assessment and risk control	Environmental aspects	Quality system
Legal and other requirements	Legal and other requirements	--
Objectives	Objectives and targets	Quality system
OH&S management program(s)	Environmental management program(s)	Quality system
Implementation and operation	Implementation and operation	Quality system Process control
Structure and responsibility	Structure and responsibility	Management responsibility Organization
Training, awareness and competence	Training, awareness and competence	Training
Consultation and communication	Communication	--
Documentation	Environmental management system documentation	General (without 1st sentence)
Document and data control	Document control	Document and data control

Inter-relationship between OHSAS 18001, ISO 14001 and ISO 9001:

OHSAS 18001	ISO 14001	ISO 9001
Operational control	Operational control	Quality system procedures Contract review Design control Purchasing Customer supplied product Product identification and traceability Process control Handling, storage, packaging, preservation and delivery Servicing Statistical techniques
Emergency preparedness and response	Emergency preparedness and response	--
Checking and corrective action	Checking and corrective action	--

Performance measurement and monitoring	Monitoring and measurement	Inspection and testing Control of inspection, measuring and test equipment Inspection and test status

| Accidents, incidents, non-conformances and corrective and preventive action | Non-conformance and corrective and preventive action | Control of non- conforming product

Corrective and preventive action |
|---|---|---|
| Records and records management | records | Control of quality records |
| Audit | Environmental management system audit | Internal quality audits |
| Management review | Management review | Management review |

BS 18004:2008

BS 18004 provides guidance for an occupational health and safety (OH&S) management system or the OH&S elements of an organization's overall management system. It enables an organization to control its OH&S risks and improve its OH&S performance.

BS 18004 applies to any organization that wishes to:

- Establish an OH&S management system to control risks to personnel and other interested parties who could be exposed to OH&S hazards associated with its activities

- Implement, maintain and continually improve an OH&S management system

- Demonstrate commitment to good practice, including selfregulation, and continual improvement in OH&S performance

- Assure itself of its conformity with its stated OH&S policy and with BS OHSAS 18001 by: Making self-determination and self-declaration, or Seeking confirmation of its conformity by parties having an interest in the organization, such as customers, or Seeking confirmation of its selfdeclaration by a party external to the organization, or Seeking certification/ registration of its OH&S management system by an external organization.

All elements of BS 18004 can be incorporated into any OH&S management system and enables an organization to incorporate OH&S within its overall management system. The extent of the application will depend on such factors as the OH&S policy of the organization, the nature of its activities and the risks and complexity of its operations.

Life-cycle Assessment

Life-cycle assessment (LCA, also known as life-cycle analysis, ecobalance, and cradle-to-grave analysis) is a technique to assess environmental impacts associated with all the stages of a product's life from raw material extraction through materials processing, manufacture, distribution, use, repair and maintenance, and disposal or recycling. Designers use this process to help critique their products. LCAs can help avoid a narrow outlook on environmental concerns by:

- Compiling an inventory of relevant energy and material inputs and environmental releases;

- Evaluating the potential impacts associated with identified inputs and releases;

- Interpreting the results to help make a more informed decision.

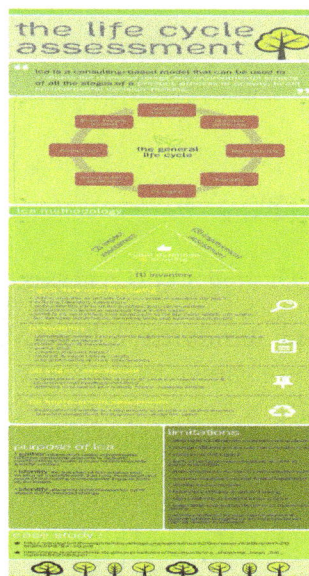

Life Cycle Assessment Overview

Goals and Purpose

The goal of LCA is to compare the full range of environmental effects assignable to products and services by quantifying all inputs and outputs of material flows and assessing how these material flows affect the environment. This information is used to improve processes, support policy and provide a sound basis for informed decisions.

The term *life cycle* refers to the notion that a fair, holistic assessment requires the assessment of raw-material production, manufacture, distribution, use and disposal including all intervening transportation steps necessary or caused by the product's existence.

There are two main types of LCA. Attributional LCAs seek to establish (or attribute) the burdens associated with the production and use of a product, or with a specific service or process, at a point in time (typically the recent past). Consequential LCAs seek to identify the environmental conse-quences of a decision or a proposed change in a system under study (oriented to the future), which means that market and economic implications of a decision may have to be taken into account. Social LCA is under development as a different approach to life cycle thinking intended to assess social implications or potential impacts. Social LCA should be considered as an approach that is complementary to environmental LCA.

The procedures of life cycle assessment (LCA) are part of the ISO 14000 environmental manage-ment standards: in ISO 14040:2006 and 14044:2006. (ISO 14044 replaced earlier versions of ISO 14041 to ISO 14043.) GHG product life cycle assessments can also comply with specifications such as PAS 2050 and the GHG Protocol Life Cycle Accounting and Reporting Standard.

Four Main Phases

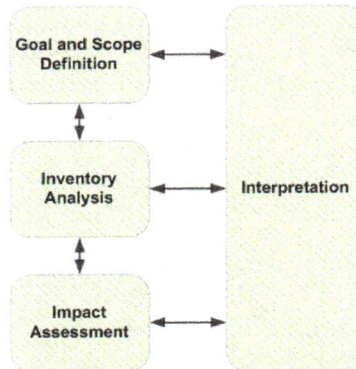

Illustration of LCA phases

According to the ISO 14040 and 14044 standards, a Life Cycle Assessment is carried out in four distinct phases as illustrated in the figure shown to the right. The phases are often interdependent in that the results of one phase will inform how other phases are completed.

Goal and Scope

An LCA starts with an explicit statement of the goal and scope of the study, which sets out the context of the study and explains how and to whom the results are to be communicated. This is a key step and the ISO standards require that the goal and scope of an LCA be clearly defined and consistent with the intended application. The goal and scope document therefore includes technical details that guide subsequent work:

- The functional unit, which defines what precisely is being studied and quantifies the service delivered by the product system, providing a reference to which the inputs and outputs can be related. Further, the functional unit is an important basis that enables alternative goods, or services, to be compared and analyzed. So to explain this a functional system which is inputs, processes and outputs contains a functional unit, that fulfills a function, for example paint is covering a wall, making a functional unit of 1m² covered for 10 years. The functional flow would be the items necessary for that function, so this would be a brush, tin of paint and the paint itself.

- The system boundaries; which are delimitations of which processes that should be included in the analysis of a product system.

- Any assumptions and limitations;

- The allocation methods used to partition the environmental load of a process when several products or functions share the same process; allocation is commonly dealt with in one of three ways: system expansion, substitution and partition. Doing this is not easy and different methods may give different results.

- The impact categories chosen for example human toxicity, smog, global warming, eutrophication.

Life Cycle Inventory

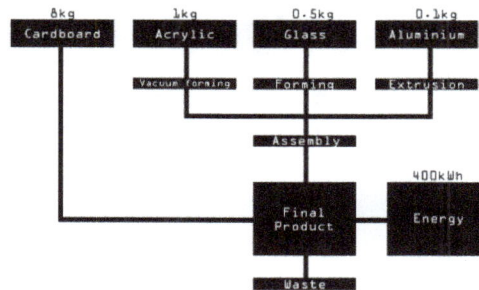

This is an example of a Life-cycle inventory (LCI) diagram

Life Cycle Inventory (LCI) analysis involves creating an inventory of flows from and to nature for a product system. Inventory flows include inputs of water, energy, and raw materials, and releases to air, land, and water. To develop the inventory, a flow model of the technical system is constructed using data on inputs and outputs. The flow model is typically illustrated with a flow chart that includes the activities that are going to be assessed in the relevant supply chain and gives a clear picture of the technical system boundaries. The input and output data needed for the construction of the model are collected for all activities within the system boundary, including from the supply chain (referred to as inputs from the technosphere).

The data must be related to the functional unit defined in the goal and scope definition. Data can be presented in tables and some interpretations can be made already at this stage. The results of the inventory is an LCI which provides information about all inputs and outputs in the form of elementary flow to and from the environment from all the unit processes involved in the study.

Inventory flows can number in the hundreds depending on the system boundary. For product LCAs at either the generic (i.e., representative industry averages) or brand-specific level, that data is typically collected through survey questionnaires. At an industry level, care has to be taken to ensure that questionnaires are completed by a representative sample of producers, leaning toward neither the best nor the worst, and fully representing any regional differences due to energy use, material sourcing or other factors. The questionnaires cover the full range of inputs and outputs, typically aiming to account for 99% of the mass of a product, 99% of the energy used in its production and any environmentally sensitive flows, even if they fall within the 1% level of inputs.

One area where data access is likely to be difficult is flows from the technosphere. The technosphere is more simply defined as the man-made world. Considered by geologists as secondary resources, these resources are in theory 100% recyclable; however, in a practical sense, the primary goal is salvage. For an LCI, these technosphere products (supply chain products) are those that have been produced by man and unfortunately those completing a questionnaire about a process which uses a man-made product as a means to an end will be unable to specify how much of a given input they use. Typically, they will not have access to data concerning inputs and outputs for previous production processes of the product. The entity undertaking the LCA must then turn to secondary sources if it does not already have that data from its own previous studies. National databases or data sets that come with LCA-practitioner tools, or that can be readily accessed, are the usual sources for that information. Care must then be taken to ensure that the secondary data source properly reflects regional or national conditions.

LCI Methods

- Process LCA

- Economic Input Output LCA

- Hybrid Approach

Life Cycle Impact Assessment

Inventory analysis is followed by impact assessment. This phase of LCA is aimed at evaluating the significance of potential environmental impacts based on the LCI flow results. Classical life cycle impact assessment (LCIA) consists of the following mandatory elements:

- selection of impact categories, category indicators, and characterization models;

- the classification stage, where the inventory parameters are sorted and assigned to specific impact categories; and

- impact measurement, where the categorized LCI flows are characterized, using one of many possible LCIA methodologies, into common equivalence units that are then summed to provide an overall impact category total.

In many LCAs, characterization concludes the LCIA analysis; this is also the last compulsory stage according to ISO 14044:2006. However, in addition to the above mandatory LCIA steps, other optional LCIA elements – normalization, grouping, and weighting – may be conducted depending on the goal and scope of the LCA study. In normalization, the results of the impact categories from the study are usually compared with the total impacts in the region of interest, the U.S. for example. Grouping consists of sorting and possibly ranking the impact categories. During weighting, the different environmental impacts are weighted relative to each other so that they can then be summed to get a single number for the total environmental impact. ISO 14044:2006 generally advises against weighting, stating that "weighting, shall not be used in LCA studies intended to be used in comparative assertions intended to be disclosed to the public". This advice is often ignored, resulting in comparisons that can reflect a high degree of subjectivity as a result of weighting.

Interpretation

Life Cycle Interpretation is a systematic technique to identify, quantify, check, and evaluate information from the results of the life cycle inventory and/or the life cycle impact assessment. The results from the inventory analysis and impact assessment are summarized during the interpretation phase. The outcome of the interpretation phase is a set of conclusions and recommendations for the study. According to ISO 14040:2006, the interpretation should include:

- identification of significant issues based on the results of the LCI and LCIA phases of an LCA;

- evaluation of the study considering completeness, sensitivity and consistency checks; and

- conclusions, limitations and recommendations.

A key purpose of performing life cycle interpretation is to determine the level of confidence in the final results and communicate them in a fair, complete, and accurate manner. Interpreting the results of an LCA is not as simple as "3 is better than 2, therefore Alternative A is the best choice"! Interpreting the results of an LCA starts with understanding the accuracy of the results, and ensuring they meet the goal of the study. This is accomplished by identifying the data elements that contribute significantly to each impact category, evaluating the sensitivity of these significant data elements, assessing the completeness and consistency of the study, and drawing conclusions and recommendations based on a clear understanding of how the LCA was conducted and the results were developed.

Reference Test

More specifically, the best alternative is the one that the LCA shows to have the least cradle-to-grave environmental negative impact on land, sea, and air resources.

LCA uses

Based on a survey of LCA practitioners carried out in 2006 LCA is mostly used to support business strategy (18%) and R&D (18%), as input to product or process design (15%), in education (13%) and for labeling or product declarations (11%). LCA will be continuously integrated into the built environment as tools such as the European ENSLIC Building project guidelines for buildings or developed and implemented, which provide practitioners guidance on methods to implement LCI data into the planning and design process.

Major corporations all over the world are either undertaking LCA in house or commissioning studies, while governments support the development of national databases to support LCA. Of particular note is the growing use of LCA for ISO Type III labels called Environmental Product Declarations, defined as "quantified environmental data for a product with pre-set categories of parameters based on the ISO 14040 series of standards, but not excluding additional environmental information". These third-party certified LCA-based labels provide an increasingly important basis for assessing the relative environmental merits of competing products. Third-party certification plays a major role in today's industry. Independent certification can show a company's dedication to safer and environmental friendlier products to customers and NGOs.

LCA also has major roles in environmental impact assessment, integrated waste management and pollution studies.

Data Analysis

A life cycle analysis is only as valid as its data; therefore, it is crucial that data used for the completion of a life cycle analysis are accurate and current. When comparing different life cycle analyses with one another, it is crucial that equivalent data are available for both products or processes in question. If one product has a much higher availability of data, it cannot be justly compared to another product which has less detailed data.

There are two basic types of LCA data – unit process data and environmental input-output data (EIO), where the latter is based on national economic input-output data. Unit process data are derived from direct surveys of companies or plants producing the product of interest, carried out at a unit process level defined by the system boundaries for the study.

Data validity is an ongoing concern for life cycle analyses. Due to globalization and the rapid pace of research and development, new materials and manufacturing methods are continually being introduced to the market. This makes it both very important and very difficult to use up-to-date information when performing an LCA. If an LCA's conclusions are to be valid, the data must be recent; however, the data-gathering process takes time. If a product and its related processes have not undergone significant revisions since the last LCA data was collected, data validity is not a problem. However, consumer electronics such as cell phones can be redesigned as often as every 9 to 12 months, creating a need for ongoing data collection.

The life cycle considered usually consists of a number of stages including: materials extraction, processing and manufacturing, product use, and product disposal. If the most environmentally harmful of these stages can be determined, then impact on the environment can be efficiently reduced by focusing on making changes for that particular phase. For example, the most energy-intensive life phase of an airplane or car is during use due to fuel consumption. One of the most effective ways to increase fuel efficiency is to decrease vehicle weight, and thus, car and airplane manufacturers can decrease environmental impact in a significant way by replacing heavier materials with lighter ones such as aluminium or carbon fiber-reinforced elements. The reduction during the use phase should be more than enough to balance additional raw material or manufacturing cost.

Data sources are typically large databases, it is not appropriate to compare two options if different data sources have been used to source the data. Data sources include:

soca

EuGeos' 15804-IA

NEEDS

ecoinvent

PSILCA

ESU World Food

GaBi

ELCD

LC-Inventories.ch

Social Hotspots

ProBas

bioenergiedat

Agribalyse

USDA

Ökobaudat

Calculations for impact can then be done by hand, but it is more usual to streamline the process by using software. This can range from a simple spreadsheet, where the user enters the data manually to a fully automated program, where the user is not aware of the source data.

Variants

Cradle-to-grave

Cradle-to-grave is the full Life Cycle Assessment from resource extraction ('cradle') to use phase and disposal phase ('grave'). For example, trees produce paper, which can be recycled into low-energy production cellulose (fiberised paper) insulation, then used as an energy-saving device in the ceiling of a home for 40 years, saving 2,000 times the fossil-fuel energy used in its production. After 40 years the cellulose fibers are replaced and the old fibers are disposed of, possibly incinerated. All inputs and outputs are considered for all the phases of the life cycle.

Cradle-to-gate

Cradle-to-gate is an assessment of a *partial* product life cycle from resource extraction (*cradle*) to the factory gate (i.e., before it is transported to the consumer). The use phase and disposal phase of the product are omitted in this case. Cradle-to-gate assessments are sometimes the basis for environmental product declarations (EPD) termed business-to-business EDPs. One of the significant uses of the cradle-to-gate approach compiles the life cycle inventory (LCI) using cradle-to-gate. This allows the LCA to collect all of the impacts leading up to resources being purchased by the facility. They can then add the steps involved in their transport to plant and manufacture process to more easily produce their own cradle-to-gate values for their products.

Cradle-to-cradle or Closed Loop Production

Cradle-to-cradle is a specific kind of cradle-to-grave assessment, where the end-of-life disposal step for the product is a recycling process. It is a method used to minimize the environmental impact of products by employing sustainable production, operation, and disposal practices and aims to incorporate social responsibility into product development. From the recycling process originate new, identical products (e.g., asphalt pavement from discarded asphalt pavement, glass bottles from collected glass bottles), or different products (e.g., glass wool insulation from collected glass bottles).

Allocation of burden for products in open loop production systems presents considerable challenges for LCA. Various methods, such as the avoided burden approach have been proposed to deal with the issues involved.

Gate-to-gate

Gate-to-gate is a partial LCA looking at only one value-added process in the entire production chain. Gate-to-gate modules may also later be linked in their appropriate production chain to form a complete cradle-to-gate evaluation.

Well-to-wheel

Well-to-wheel is the specific LCA used for transport fuels and vehicles. The analysis is often broken down into stages entitled "well-to-station", or "well-to-tank", and "station-to-wheel" or "tank-to-wheel", or "plug-to-wheel". The first stage, which incorporates the feedstock or fuel production and processing and fuel delivery or energy transmission, and is called the "upstream" stage, while the stage that deals with vehicle operation itself is sometimes called the "downstream" stage. The well-to-wheel analysis is commonly used to assess total energy consumption, or the energy conversion efficiency and emissions impact of marine vessels, aircraft and motor vehicles, including their carbon footprint, and the fuels used in each of these transport modes.

The well-to-wheel variant has a significant input on a model developed by the Argonne National Laboratory. The Greenhouse gases, Regulated Emissions, and Energy use in Transportation (GREET) model was developed to evaluate the impacts of new fuels and vehicle technologies. The model evaluates the impacts of fuel use using a well-to-wheel evaluation while a traditional cradle-to-grave approach is used to determine the impacts from the vehicle itself. The model reports energy use, greenhouse gas emissions, and six additional pollutants: volatile organic compounds (VOCs), carbon monoxide (CO), nitrogen oxide (NOx), particulate matter with size smaller than 10 micrometre (PM10), particulate matter with size smaller than 2.5 micrometre (PM2.5), and sulfur oxides (SOx).

Economic Input–output Life Cycle Assessment

Economic input–output LCA (EIOLCA) involves use of aggregate sector-level data on how much environmental impact can be attributed to each sector of the economy and how much each sector purchases from other sectors. Such analysis can account for long chains (for example, building an automobile requires energy, but producing energy requires vehicles, and building those vehicles requires energy, etc.), which somewhat alleviates the scoping problem of process LCA; however, EIOLCA relies on sector-level averages that may or may not be representative of the specific subset of the sector relevant to a particular product and therefore is not suitable for evaluating the environmental impacts of products. Additionally the translation of economic quantities into environmental impacts is not validated.

Ecologically Based LCA

While a conventional LCA uses many of the same approaches and strategies as an Eco-LCA, the latter considers a much broader range of ecological impacts. It was designed to provide a guide to wise management of human activities by understanding the direct and indirect impacts on ecological resources and surrounding ecosystems. Developed by Ohio State University Center for resilience, Eco-LCA is a methodology that quantitatively takes into account regulating and supporting services during the life cycle of economic goods and products. In this approach services are categorized in four main groups: supporting, regulating, provisioning and cultural services.

Exergy Based LCA

Exergy of a system is the maximum useful work possible during a process that brings the system into equilibrium with a heat reservoir. Wall clearly states the relation between exergy analysis and resource accounting. This intuition confirmed by DeWulf and Sciubba lead to Exergo-economic accounting and to methods specifically dedicated to LCA such as Exergetic material input per unit of service (EMIPS). The concept of material input per unit of service (MIPS) is quantified in terms of the second law of thermodynamics, allowing the calculation of both resource input and service output in exergy terms. This exergetic material input per unit of service (EMIPS) has been elaborated for transport technology. The service not only takes into account the total mass to be transported and the total distance, but also the mass per single transport and the delivery time.

The applicability of the EMIPS methodology relates specifically to transport system. This model has been further improved by Trancossi who has introduced the friction term, which has not been considered by original EMIPS model, and the key distinction between exergy disruption by payload and by vehicle, focusing on the losses due to vehicle and more effective evaluation of the processes and produced an effective assessment of today transport vehicles. This model is referenced by Indian "Road less traveled" model, which has been developed for minimizing the impact of transports in urban environment.

Life Cycle Energy Analysis

Life cycle energy analysis (LCEA) is an approach in which all energy inputs to a product are accounted for, not only direct energy inputs during manufacture, but also all energy inputs needed to produce components, materials and services needed for the manufacturing process. An earlier term for the approach was *energy analysis*.

With LCEA, the *total life cycle energy input* is established.

Energy Production

It is recognized that much energy is lost in the production of energy commodities themselves, such as nuclear energy, photovoltaic electricity or high-quality petroleum products. *Net energy content* is the energy content of the product minus energy input used during extraction and conversion, directly or indirectly. A controversial early result of LCEA claimed that manufacturing solar cells requires more energy than can be recovered in using the solar cell. The result was refuted. Another new concept that flows from life cycle assessments is Energy Cannibalism. Energy Cannibalism refers to an effect where rapid growth of an entire energy-intensive industry creates a need for energy that uses (or cannibalizes) the energy of existing power plants. Thus during rapid growth the industry as a whole produces no energy because new energy is used to fuel the embodied energy of future power plants. Work has been undertaken in the UK to determine the life cycle energy (alongside full LCA) impacts of a number of renewable technologies.

Energy Recovery

If materials are incinerated during the disposal process, the energy released during burning can be harnessed and used for electricity production. This provides a low-impact energy source, especially when compared with coal and natural gas While incineration produces more greenhouse gas emissions than landfilling, the waste plants are well-fitted with filters to minimize this negative impact. A recent study comparing energy consumption and greenhouse gas emissions from landfilling (without energy recovery) against incineration (with energy recovery) found incineration to be superior in all cases except for when landfill gas is recovered for electricity production.

Criticism

A criticism of LCEA is that it attempts to eliminate monetary cost analysis, that is replace the currency by which economic decisions are made with an energy currency. It has also been argued that energy efficiency is only one consideration in deciding which alternative process to employ, and that it should not be elevated to the only criterion for determining environmental acceptability; for example, simple energy analysis does not take into account the renewability of energy flows or the toxicity of waste products; however the life cycle assessment does help companies become more familiar with environmental properties and improve their environmental system. Incorporating Dynamic LCAs of renewable energy technologies (using sensitivity analyses to project future improvements in renewable systems and their share of the power grid) may help mitigate this criticism.

In recent years, the literature on life cycle assessment of energy technology has begun to reflect the interactions between the current electrical grid and future energy technology. Some papers have focused on energy life cycle, while others have focused on carbon dioxide (CO_2) and other greenhouse gases. The essential critique given by these sources is that when considering energy technology, the growing nature of the power grid must be taken into consideration. If this is not done, a given class of energy technology may emit more CO_2 over its lifetime than it mitigates.

A problem the energy analysis method cannot resolve is that different energy forms (heat, electricity, chemical energy etc.) have different quality and value even in natural sciences, as a consequence of the two main laws of thermodynamics. A thermodynamic measure of the quality of energy is exergy. According to the first law of thermodynamics, all energy inputs should be accounted with equal weight, whereas by the second law diverse energy forms should be accounted by different values.

The conflict is resolved in one of these ways:

- value difference between energy inputs is ignored,

- a value ratio is arbitrarily assigned (e.g., a joule of electricity is 2.6 times more valuable than a joule of heat or fuel input),

- the analysis is supplemented by economic (monetary) cost analysis,

- exergy instead of energy can be the metric used for the life cycle analysis.

Critiques

Life cycle assessment is a powerful tool for analyzing commensurable aspects of quantifiable systems. Not every factor, however, can be reduced to a number and inserted into a model. Rigid system boundaries make accounting for changes in the system difficult. This is sometimes referred to as the boundary critique to systems thinking. The accuracy and availability of data can also contribute to inaccuracy. For instance, data from generic processes may be based on averages, unrepresentative sampling, or outdated results. Additionally, social implications of products are generally lacking in LCAs. Comparative life-cycle analysis is often used to determine a better process or product to use. However, because of aspects like differing system boundaries, different statistical information, different product uses, etc., these studies can easily be swayed in favor of one product or process over another in one study and the opposite in another study based on varying parameters and different available data. There are guidelines to help reduce such conflicts in results but the method still provides a lot of room for the researcher to decide what is important, how the product is typically manufactured, and how it is typically used.

An in-depth review of 13 LCA studies of wood and paper products found a lack of consistency in the methods and assumptions used to track carbon during the product lifecycle. A wide variety of methods and assumptions were used, leading to different and potentially contrary conclusions – particularly with regard to carbon sequestration and methane generation in landfills and with carbon accounting during forest growth and product use.

Streamline LCA

This process includes three steps. First, a proper method should be selected to combine adequate accuracy with acceptable cost burden in order to guide decision making. Actually, in LCA process, besides streamline LCA, Eco-screening and complete LCA are usually considered as well. However, the former one only could provide limited details and the latter one with more detailed information is more expensive. Second, single measure of stress should be selected. Typical LCA output includes resource consumption, energy consumption, water consumption, emission of CO_2, toxic residues and so on. One of these outputs is used as the main factor to measure in streamline LCA. Energy consumption and CO_2 emission are often regarded as "practical indicators". Last, stress selected in step 2 is used as standard to assess phase of life separately and identify the most damaging phase. For instance, for a family car, energy consumption could be used as the single stress factor to assess each phase of life. The result shows that the most energy intensive phase for a family car is the usage stage.

Life Cycle Assessment of Engineered Material in Service plays a significant role in saving energy, conserving resources and saving billions by preventing premature failure of critical engineered component in a machine or equipment. LCA data of surface engineered materials are used to improve life cycle of the engineered component. Life cycle improvement of industrial machineries and equipments including, manufacturing, power generation, transportations, etc. leads to improvement in energy efficiency, sustainability and negating global temperature rise. Estimated reduction in anthropogenic carbon emission is minimum 10% of the global emission.

Life Cycle Assessment and its Purpose

Life cycle assessment (LCA) is a tool to evaluate the environmental effects of a product or process throughout its entire life cycle. An LCA entails examining the product from the extraction of raw materials for the manufacturing process, through the production and use of the item, to its final disposal, and thus encompassing the entire product system.

Product Life Cycle

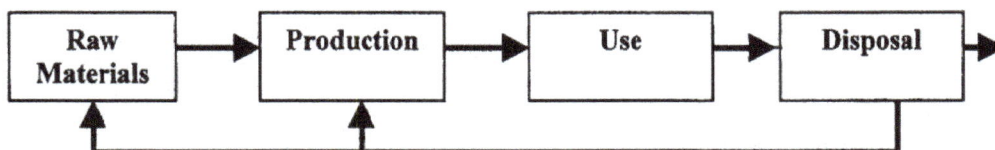

The assessment process includes identifying and quantifying energy and materials used and wastes released to the environment, assessing their environmental impact and evaluating opportunities for improvement.

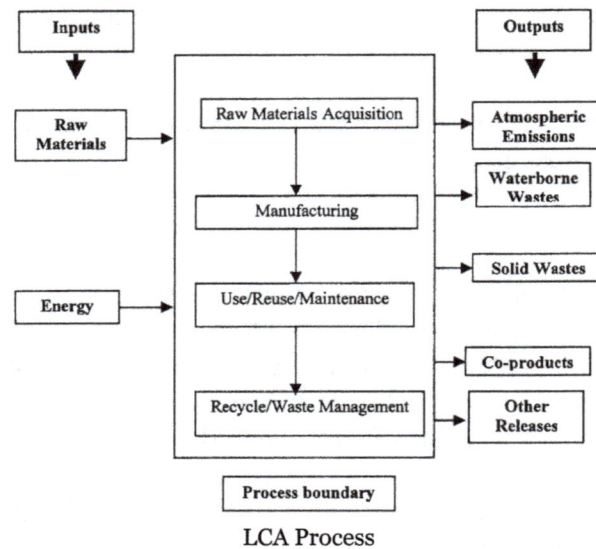

LCA Process

The unique feature of this type of assessment is its focus on the entire life cycle, rather than a single manufacturing step or environmental emission. The theory behind this approach is that operations occurring within a facility can also cause impacts outside the facility's gates that need to be considered when evaluating project alternatives.

Purpose of LCA

Government and customers who purchase products from different sectored companies are keen on environmental properties of all the products. EMAS and ISO 14000 are demanding continual improvement in the process of production and in the environmental management system. At this stage there is a need for a tool like LCA that helps organisations meet the demand to improvise process/product.

Evolution oF Life Cycle Assessment

The principles underlying an LCA were developed in the late 1960s. In the 1970's, the US Environmental protection agency refined the methodology for evaluation of environmental impacts of products and were popularly known as resource and environmental profile analysis (REPA). Initially, it was used mainly on the consumption of energy and other resources. Knowledge of environmentally damaging releases and actions and the estimation of their effects, was too rudimentary at that time to allow a quantitative treatment of the environmental impacts of the product life cycle.

Assessments of product life cycle experienced a renaissance through studies of the environmental loadings and potential impacts of beverage containers (e.g., beer cans, milk containers) performed in various European countries in the early 1980s. These studies involved further elaborations of the principles underlying the assessment of product life cycle and entailed a series of life cycle assessments of materials used in packaging containers (i.e., polyethylene, cardboard, aluminium, etc.). A common feature of the items analysed was their homogenous character and their widespread use in many different contexts.

The late 1980s and early 1990s have seen international attempts to standardise the principles underlying life cycle assessments and to develop codes of good conduct in this field. The list of products that have been subjected to LCA has grown quickly and now includes more complex products such as paints, insulation materials, window frames, refrigerators, hotplates, television sets, etc., as well as the entire service systems or technologies such as electricity production.

As a part of ISO 14000 series of standards ISO in 2000 has come out with the following standards:

- ISO 14040: Environmental Management – LCA – Principles and Framework.

- ISO 14041: Environmental Management – LCA – Inventory Analysis.

- ISO 14042: Environmental Management – LCA – Impact Assessment.

- ISO 14043: Environmental Management – LCA – Interpretation.

Since the last decade or so, LCA is gaining importance as an environmental management tool. It has now emerged as a decision support tool in such areas as business, regulation and policy and to structure technology development in a coherent way. Many potential applications of LCA are envisaged including product improvement and design, environmental management, eco-labelling, green accounting, environmental auditing and reporting, resource management, definition of best available technology (BAT), product policy, strategic industrial planning, strategic environmental policy development, etc. As a general concept, the life cycle approach aims to support the overall goal of sustainability.

Stages in Product LCA

LCA is split into five stages that include:

1. Planning : Includes Statement of objectives, Definition of the product and its alternatives, Choice of system boundaries, Choice of environmental parameters, Choice of aggregation and evaluation method and Strategy for data collection

2. Screening: Includes preliminary execution of LCA and adjustment of plan

3. Data collection and treatment: Includes measurements, interviews, literature search, theoretical calculations, database search, qualified guessing and also computation of the inventory table

4. Evaluation: Includes classification of inventory table into impact categories, aggregation within category, normalisation and weighting of different categories

5. Improvement assessment: Includes sensitivity analysis and improvement priority and feasibility assessment

It is generally recognized that the first stage is extremely important. The result of the LCA is heavily dependent on the decisions taken in this phase.

The screening LCA is a useful step to check the goal-definition phase. After screening it is much easier to plan the rest of the project.

The following figure gives a diagrammatic representation of the stages in the life cycle of a product:

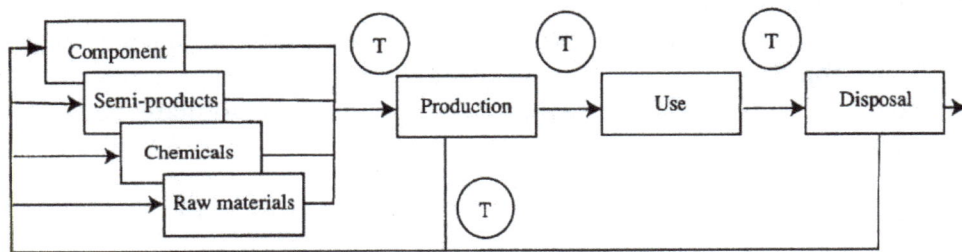

T = Transportation

Product Life Cycle Stages

As depicted in these figures, each stage of the life cycle receives materials and energy as inputs and produces:

- outputs of material or products to subsequent life cycle stages;

- emissions to the environment.

There are a number of issues associated with the life cycle stages, and we will touch upon some of these, next.

Extraction of Raw Materials

This stage in the life cycle includes the extraction of all materials involved in the entire life cycle of the product. Typical examples of activities included in this stage are forest logging and crop harvesting, fishing and mining of ores and minerals. The inventory for the extraction of raw materials should include raw materials for the production of the machinery (i.e., capital equipment) involved in manufacturing the product and other stages of the product life cycle. Raw materials used in the production of electricity and energy used in the different life stages of the product should also be considered. Collecting data for the raw materials extraction stage may prove to be a complex task. It may also lead into iterative processes such as assessing the inputs and outputs related to extraction of the raw materials that is used in the production of end products. Often, the most serious environmental problems of the product life cycle associated with this first stage. It is a common error to leave out parts of the raw materials stage from the LCA. Essentially, the decision of what to include or exclude in the LCA should be based on a sensitivity analysis.

Manufacture of a Product

The manufacturing stage encompasses all the processes involved in the conversion of raw materials into the products considered in the LCA. Apart from the manufacturing processes at the plant where the product is made, this stage takes into account production of ancillary materials, chemicals and specific or general components at other plants, no matter where they are.

Transportation

Transportation is really not a single life stage in itself. Rather, it is an integral part of all stages of the life cycle. Transportation could be characterised as conveyance of materials or energy between different operations at various locations. Included in this stage, apart from the transport process itself, is the production of packaging materials for the transportation of the product. The transport stage would possibly also include an appropriate share of the environmental loadings and consumptions associated with the construction and maintenance of the transport system, whether this is road, rail, water or air transportation.

Use of Product

The use-stage of the product occurs when it is put in service and operated over its useful life. This begins after the distribution of the product and ends when the product is used up or discarded to the waste management system. Included in the use-stage are releases and resource consumptions created by the use or maintenance of the product.

Waste Management

Wastes are generated in each phase of the life cycle, and they need to be properly managed to protect the environment. The management of wastes may involve alternative processes such as the following:

(i) Reuse: This means the use of the product or parts thereof in new units of the same product or in different products.

(ii) Recycling: This means the use of materials in the product for manufacture of the same or other products.

(iii) Incineration: This refers to the combustion of the product, generating heat that may be used for electricity production or heating.

(iv) Composting: This refers to the microbial degradation of biological materials yielding compost for improvement of agricultural soils.

(v) Waste water treatment: This refers to the organic matter degradation and nutrients removal from sewage water, creating sludge that is deposited on agricultural land.

(vi) Land filling: This means the deposition of the product in landfills.

Each form of waste treatment mentioned above may be considered a processing of waste associated with a certain consumption of resources. This results in various releases into the environment, and the possible generation of energy or materials that will be an input to the manufacturing process of this product or of other products. Before you read any further, it is a good idea to look at the Course Municipal Solid Waste Management in which we discussed the different waste management processes in detail.

As implied, the LCA of products is an important environmental management tool. However, as with every tool, difficulties do arise with LCA too.

A Code of Good Conduct for LCA

As mentioned earlier, LCA emerged during the last decade as a tool to provide an objective assessment of the total environmental impact associated with a product through its entire life cycle. In several countries, LCA is considered the primary tool by which environmental impacts of products should be regulated by government authorities. However, problems do exist. Let us look at some of the major LCA limitations below, as revealed in various studies:

(i) Data quality: In a manufacturer-sponsored study to compare a product with its alternatives, the consultants performing the LCA were able to get a very detailed and current data from the manufacturer for the processes involved in the production of that product. However, they had to depend on secondary data from the literature or earlier studies with regard to the production processes concerning the alternative products. Obviously, comparative studies on the basis of the secondary data tend to lack credibility.

(ii) Life cycle boundaries: A Dutch study excluded the production of several raw materials, including crude oil, for the polycarbonate production. The German study did not include emissions from the energy production associated with the life cycle of the milk containers. Most studies did not consider the working environment.

(iii) Country-specific technology types: An example is the LCA conducted on the production of electricity used in the product life cycle. In a Swedish study, the electricity production was based on nuclear power and hydroelectricity, while a Swiss study, based on a US energy scenario since 1972, used coal as the energy source. This should explain the difference in the emissions found in the two studies. In addition, as the waste processing systems contribute to the nature and level of emissions, it is important to examine these systems as well. For example, if the waste is incinerated and the combustion energy is used for electricity production, the Dutch study may tend to favour carton containers over recyclable polycarbonate bottles.

(iv) Evaluation stage priorities: In the cases where alternatives do not have particular advantage over one another, the priorities in the evaluation stage become decisive for the outcome. A recommendation of the Danish study (referred to at (ii) above) was based on an evaluation stage choice between reduced consumption of raw materials and water, but on the releases of dichloromethane into the working environment (for polycarbonate bottles), and higher energy consumption and loading of bio-accumulating and possibly carcinogenic chlorinated compounds in the waste from production (for the milk cartons).

 Considering the versatility and many diverse possible applications of the LCA tool, it may be difficult to obtain reproducible and consistent results through standardisation, without losing the necessary flexibility for adaptation to particular cases being studied. Given this scenario, it is important that we include the following factors in a code of good conduct for LCA:

(v) Definition: This involves finding answers to such questions as: Is the purpose of the study explicitly defined? Is it meant for internal company use or for public use? If the study is intended for public use, has it been peer reviewed? Is it clearly stated for whom the study is performed and by whom it is sponsored? Is the definition of the functional unit appropriate?, etc.

(vi) Delimitation of system under study: This involves finding answers to such questions as: Is there an explicit and clear delimitation of the system under study? Is the life cycle described in detail, stage by stage? Are the life cycle description and process tree plausible? Do they describe the real world system in a realistic way? Does the study include the extraction of raw materials? Does the study include the production of electricity? What production scenario is used? Is it appropriate? Does the study include the manufacture of real capital for all life cycle stages? If the study has precluded capital, is this omission substantiated? Do you find the omission reasonable? Is the disposal stage covered by the study?, etc.

(vii) Inventory: This involves finding answers to such questions as: Does the inventory cover all processes of the process tree? Is there a reference to the source of every piece of data in the inventory? Is the data quality appropriate, i.e., primary and recent data for all the important processes of the life cycle? Do the data describe a relevant technological level of the processes? Is the use of data of lower quality or omission of processes from the inventory based on sensitivity analyses? Is the quality of the data used for the compared alternatives similar? Are data on impacts to the working environment present in the inventory? Is it acceptable that they be left out in the considered study?, etc.

(viii) Impact assessment: This involves finding answers to such questions as: Has any impact assessment been performed? Does it consider all the important environmental effect types? Does it consider resource and working environment issues?, etc.

(ix) Evaluation: This involves finding answers to such questions as: Are the subjective steps of the evaluation separated from the objective ones? If not, is the evaluation transparent? Are the priorities clear? Are the conclusions of the assessment clear? Are the assumptions underlying the weighting explained, and do you agree with them?, etc.

Procedure for LCA

Four steps are involved in carrying out an LCA, and these are:

(i) Definition of scope, goals, and delimitation of the life cycle.

(ii) Preparation of an inventory.

 (iii) Assessment of impact of environmental loadings in terms of environmental profiles.

(iv) Evaluation of environmental profiles according to the defined goals.

Defining the Goal and Scope

The goal of an LCA study must unambiguously state the intended application, including the reasons for carrying out the study and the intended audience. The statement of goal must also indicate the intended use of the results and users of the results. The practitioner, who has to reach the goal, needs to understand the detailed purpose of the study in order to make proper decisions throughout the study. Examples of goals of life cycle assessments are:

- To compare two or more different products fulfilling the same function with the purpose of using the information in marketing of the products or regulating the use of the products.

- To identify improvement possibilities in further development of existing products or in innovation and design of new products.

- To identify areas, steps, etc., in the life cycle of a product to meet the eco-labelling criteria.

Transparency is essential for all kinds of LCA studies. The target group of the LCA study is also important to have in mind the choice of reporting method. The goal can be redefined as a result of the findings throughout the study.

While the goal definition determines the level of sophistication of the study and the requirements to reporting, the definition of the scope of an LCA sets the borders of the assessment, i.e., what may be included in the system and what detailed assessment methods are to be used.

In defining the scope of an LCA study, the following elements must be considered and clearly described:

- The functions of the system, or in the case of comparative studies, systems.
- The functional unit.
- The system to be studied.
- The system boundaries.
- The allocation procedures.
- The types of impact and the methodology of impact assessment and subsequent interpretation to be used.
- Data requirements.
- Assumptions.
- Limitations.
- The initial data quality requirements.
- The type of critical review, if any.
- The type and format of the report required for the study

The scope should be sufficiently well defined to ensure that the breadth, the depth and the detail of the study are compatible and sufficient to address the stated goal. LCA is an iterative technique.

Therefore, the scope of the study may need to be modified while the study is being conducted as additional information is collected.

Let us now describe some of the elements mentioned above to define the scope of an LCA.

Functional Unit

Definition of the functional unit or performance characteristics is the foundation of an LCA because the functional unit sets the scale for comparison of two or more products including improvement to one product (system). All data collected in the inventory phase will be related to the functional unit. When comparing different products fulfilling the same function, definition of the functional unit is of particular importance. The three aspects that have to be taken into account when defining a functional unit are the efficiency of the product, the durability of the product and the performance quality standard (Lindfors et al., 1995c).

When performing an assessment of more complicated systems, e.g., multi-functional systems, special attention has to be paid to by-products. Waste treatment systems are an example of processes with different outputs (e.g., energy, a fertiliser). When comparing different systems, inclusion of the produced amount of energy and fertiliser is an example of handling of different byproducts in the definition of the functional unit.

System Boundaries

The system boundaries define the processes/operation (e.g., manufacturing, transport and waste management processes) and the inputs and outputs to be taken into account in the LCA. The input can be the overall input to a production as well as input to a single process, and the same is true for the output. The definition of system boundaries is a quite subjective operation and includes geographical boundaries, life cycle boundaries (i.e., limitations in the life cycle) and boundaries between the technosphere and biosphere. Due to the subjectivity of definition of system boundaries, transparency of the defining process and the assumptions are extremely important. Note that wastewater treatment is an example of a process that often is omitted when defining the system boundaries.

Data Quality

The quality of the data used in the life cycle inventory is naturally reflected in the quality of the final LCA. The data quality can be described and assessed in different ways. It is important that the data quality is described and assessed in systematic ways that allows others to understand and control the actual data quality. Further descriptions, which define the nature of the data collected from specific sites and that from published sources and indicate whether the data should be measured, calculated or estimated, should also be considered for maintaining data quality.

n all studies, however, the following additional data quality indicators are to be taken into consideration at a certain level of detail, depending on goal and scope definition:

- Precision: That is, the measure of variability of the data values for each category expressed (e.g., variance).

- Completeness: That is, the percentage of locations reporting primary data from the potential number in existence for each data category in a unit process.

- Representativeness: This means the qualitative assessment of the degree to which the data set reflects the true population of interest (i.e., geographic, time period and technology coverage).

- Reproducibility: This means the qualitative assessment of the extent to which formation about the methodology and data values allows an independent practitioner to reproduce the results reported in the study.

- Consistency: This means the qualitative assessment of how uniformly the study methodology is applied to the various components of the analysis.

Critical Review Process

The purpose of the critical review process is to ensure the quality of the life cycle assessment. The review can be either internal or external. This may also involve the interested parties as defined within the goal and scoping definition. The critical review process ensures that:

- the methods used to carry out the LCA are consistent with the international standard and are scientifically and technically valid;

- the data used are appropriate and reasonable in relation to the goal of the study;

- the interpretations reflect the limitations identified and the goal of the study;

- the study report is transparent and consistent.

Analysing the Inventory

Inventory analysis is the second phase in an LCA, consisting of issues such as data collection, refining system boundaries, calculation, verification of data, relating data to the specific system and allocation.

We will describe each of these issues, next.

Data Collection

The inventory analysis includes collection and treatment of data to be used in preparation of a material consumption, waste and emission profile for the phases in each life cycle. The data can be both site-specific (e.g., from specific companies, specific areas and specific countries) and general sources (e.g., trade organisations, public surveys, etc.) The data have to be collected from all single processes in the life cycle. The quantitative data are important in comparisons of processes or materials, but often the quantitative data are missing or the quality is poor (too old or not technically representative, etc.). However, a more descriptive qualitative data can be used for environmental aspects or single steps in the life cycle that cannot be quantified. This can be used when the goal and scope definitions allow a non-quantitative description of the conditions.

Data collection is often the most work intensive part of a life cycle assessment, especially if site-specific data are required for all the single processes in the life cycle. In many cases, average data from the literature (often previous investigations of the same or similar products or materials) or data from trade organisations are used. The average data can be used in the conception or simplified LCA to get a first impression of the potential inputs and outputs from producing specific materials. When doing a detailed LCA, site-specific data is preferred. Note that since average data are often some years old, they may not represent the latest in technological development.

Refining System Boundaries

The system boundaries are defined as part of the scope definition procedure. After the initial data collection, the system boundaries can be refined, i.e., as a result of decisions of exclusion of life stages or sub-systems, exclusion of material flows or inclusion of new unit processes shown to be significant, according to the sensitivity analysis. The results of this refining process and the sensitivity analysis are to be documented. This analysis serves to limit the subsequent data handling to those inputs and output data, which are determined to be significant to the goal of the LCA study.

Calculation Procedures

No formal demand exists for calculation in life cycle assessment except the described demands for allocation procedures. Due to the amount of data, it is recommended, as a minimum, to develop a spreadsheet for the specific purpose. The appropriate programme can be chosen, depending on the kind and amount of data to be handled.

Validation of Data

The validation of data has to be conducted during the data collection process in order to improve the overall data quality. Systematic data validation may point out the areas where data quality must be improved or data must be found in the similar processes. During the process of data collection, a permanent and iterative check on data validity should be conducted. Validation may involve establishing mass balances, energy balances and/or comparative analyses of emission factors. Obvious anomalies in the data appearing from such validations will result in alternative data values complying with the data quality requirements.

For each data category and for each reporting location where missing data are identified, the treatment of the missing data should result in an acceptable reported data value, a zero data value if justified and a calculated value, based on the reported value from unit processes employing similar technology.

Relating Data

The fundamental input and output data are often delivered from industry in arbitrary units, e.g., energy consumption as MJ/machine/week emissions to the sewage system as mg metals/liter wastewater. The specific machine or wastewater stream is rarely connected to the production of the considered product alone but often to a number of similar products or perhaps to the whole production activity.

For each unit process, an appropriate reference flow has to be determined (e.g., one kilogram of material, one mega joule for energy, etc.). The quantitative input and output data of the unit process shall be calculated in relation to this reference flow. Based on the refined flow chart and systems boundary, unit processes are interconnected to allow calculations of the complete system. The calculation should result in all system input and output data being referenced to the functional unit. Care should be taken when aggregating the inputs and the outputs in the product system. The level of aggregation should be sufficient to satisfy the goal of the study.

Allocation and Recycling

When performing a life cycle assessment of a complex system, it may not be possible to handle all the impacts and outputs inside the system boundaries. This problem can be solved through either of the following ways:

- expanding the system boundaries to include all the inputs and outputs;
- allocating the relevant environmental impacts to the studied system.

When avoiding allocation by expanding the system boundaries, there is a risk of making the system too complex. The data collection, impact assessment and interpretation can then become too expensive and unrealistic in terms of time and money. Allocation may be a better alternative, if an appropriate method can be found for solving the actual problem. Allocation can be necessary when dealing with:

- multi-output black box processes, i.e., when more than one product is produced and some of those product flows are crossing the system boundaries;

- multi-input processes, such as waste treatment, where a strict quantitative causality between inputs and emissions seldom exists;

- open-loop recycling, where a waste material leaving the system boundaries is used as a raw material by another system, outside the boundaries of the studied system.

On the basis of the principles presented above, the following descending order of allocation procedures is generally recommended:

- Wherever possible, allocation should be avoided or minimised. This may be achieved by subdividing the unit process into two or more sub-processes, some of which can be excluded from the system under study. Transport and materials handling are examples of processes, which can sometimes be partitioned in this way.

- Where allocation cannot be avoided, the system inputs and outputs should be partitioned among their different products or functions in such a way as to reflect the underlying physical relationship among them. That is to say, they must reflect the way in which the inputs and outputs are changed by quantitative changes in the products or functions delivered by the system.

- Where physical relationship cannot be established or used as the basis for allocation, the inputs should be allocated between the products and functions in such a way as to reflect economic relationships between them. For example, burdens might be allocated between co-products in proportion to the economic value of the products.

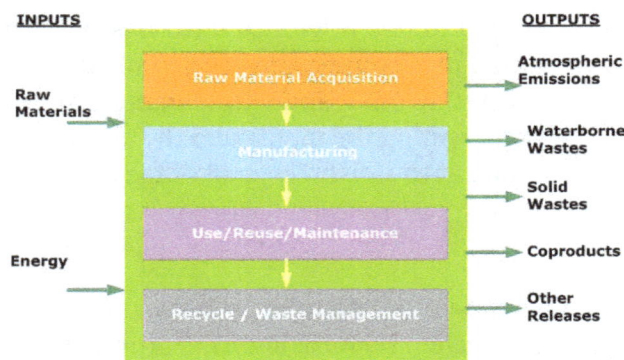

Life Cycle Stages (Source: EPA, 2001)

Assessing Environmental Impact

Impact assessment involves category definition, classification, characterisation and valuation/weighting. Let us now discuss each of these elements.

Category Definition

The life cycle assessment involves, as a first element, the definition of the impact categories to be considered (ISO, 1997c). This is a follow-up of the decisions made in the goal and scoping phase. Based on the type of information collected in the inventory phase, however, the boundaries defined in the goal and scoping may be redefined.

The impact categories are selected in order to describe the impacts caused by the products or product systems considered. The issues that need to be considered when selecting impact categories include the following:

- Completeness: This means that all environmental problems of relevance should be covered by the list.

- Practicality: This means that the list should not contain too many categories.

- Independence: This means that double counting should be avoided by choosing mutually independent impact categories.

- Relation to the characterisation step: This means that the chosen impact categories should be related to available characterisation methods (Lindfors et. al., 1995).

The impact categories considered are abiotic resources, biotic resources, land use, global warming, stratospheric ozone depletion, ecotoxicological impacts, human toxicological impacts, photochemical oxidant formation, acidification, eutrophication and work environment.

Classification

The life cycle impact assessment includes, as a second element, classification of the inventory input and output data (ISO, 1997c).

Classification is a qualitative step based on a scientific analysis of the relevant environmental processes. The classification has to assign the inventory input and output data to potential environmental impacts, i.e., impact categories. Some outputs contribute to more than one impact category, and therefore, they have to be mentioned twice. The resulting double counting is acceptable if the effects are independent of each other. However, double counting of different effects in the same effect chain (e.g., stratospheric ozone depletion and human toxicological effects, such as skin cancer) is not allowed.

The impact categories can be placed on a scale dividing the categories into four different space groups: global impacts, continental impacts, regional impacts and local impacts. The grouping is not mutually exclusive (e.g., environmental toxicity can be global, continental, regional as well as local.) The impact categories are often related directly to exposure (e.g., global exposure leads to continental impacts). Some of the impact categories are strongly correlated with continental, regional or local conditions. Certain lakes in Scandinavia can be mentioned as examples of localities that are more predisposed to acidification than lakes in other parts of Europe. The time aspect is also important when considering certain impact categories (e.g., global warming and stratospheric ozone depletion with time horizons of 20 to 500 years).

Now, let us consider an example of classification. In the manufacture of refrigerators, many of the activities relating to the production process involve the combustion of fossil fuels (e.g., coal, oil and natural gas) for electricity or heat generation, and combustion causes various emissions that may be classified as:

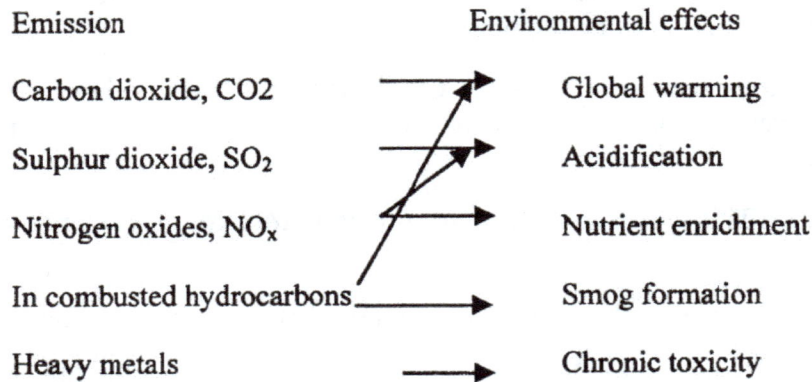

Emission	Environmental effects
Carbon dioxide, CO_2	Global warming
Sulphur dioxide, SO_2	Acidification
Nitrogen oxides, NO_x	Nutrient enrichment
In combusted hydrocarbons	Smog formation
Heavy metals	Chronic toxicity

Classification of Emissions

Characterisation

The life cycle impact assessment includes, as a third element, the characterisation of the inventory data (ISO, 1997 c).

Characterisation is mainly a quantitative step, based on scientific analyses of the relevant environmental processes. It has to assign the relative contribution of each input and output to the selected impact categories. The potential contribution of each input and output to the environmental impacts has to be estimated. For some of the environmental impact categories, there is consensus about equivalency factors to be used in the estimation of the total impact (e.g., global warming potentials, ozone depletion potentials etc.) while for some others there may not be any consensus (e.g., biotic resources, land uses, etc.).

Valuation/Weighting

Characterisation results in a quantitative statement on different impact categories (e.g., global warming, stratospheric ozone depletion and ecotoxicological effects.) Comparison of these categories is not immediately possible. Therefore, the life cycle assessment includes, as a fourth element, a valuation/weighting of the impact categories against each other.

Weighting aims to rank, weight, or if possible, aggregate the results of different life cycle impact assessment categories in order to arrive at the relative importance of these different results. The weighting process is not technical, scientific or objective, as these various life cycle impact assessment results are not directly comparable (e.g., indicators for greenhouse gases or resource depletion.) However, applying scientifically based analytical techniques may assist in weighting. The purpose of weighting the following aspects:

- To express the relative preference of an organisation or group of stakeholders based on policies, goals or aims and personal or group opinions or beliefs common to the group.

- To ensure that process is visible, documentable and reportable.

- To establish the relative importance of the results is based on the state of knowledge about these issues.

Different institutions based on different principles have developed weighting approaches including the following:

- Proxy approach: In this approach, one or a combination of several quantitative measures is stated to be indicative of the total environmental impact. Energy consumption, material displacement and space consumption are examples of using this approach.

- Technology abatement approach: The possibility of reducing environmental burdens by using different technological abatement methods can be used to set a value on the specific environmental burden. This approach can be applied to inventory data as well as impact scores.

- Monetarisation: This approach is based on the premise that utilitarianism (values are measured by the aggregation of human preferences), willingness to pay/accept is an adequate measure of preferences and the values of environmental quality can be substituted by other commodities. This approach can be applied to inventory data as well as impact scores.

- Authorised goals or standards: Environmental standards and quality targets as well as political reduction targets can be used to calculate critical volumes for emissions to air, water, soil or work environment. National or local authorities within a company can formulate the targets or standards.

- Authoritative panels: An authoritative panel can be made up of lay people, experts, representatives of governments, international bodies, etc. The credibility of a panel can be improved by using LCA experts from different societal groups as panellists, peer review sets of valuation criteria and the rules for their application, and a transparent ranking technique, and the documentation of the arguments leading to the final valuation (Volkvein, et. al., 1996).

In the impact assessment step of an LCA, the inventory of chemical and physical releases is translated into an environmental profile and the resources consumed are divided into renewable and non-renewable.

Quantitative Inventories of Environmental Releases and Resources Consumptions for the Production of 1 MJ Heat from Various Fossil Fuels:

Emission type	Unit	Coal	Heavy fuel	Natural gas	Wood	Straw
Carbon dioxide, CO_2	g/MJ	96	80	56	98a	99a
Carbon monoxide, CO	mg/MJ	50	10	150	6000	4500
Sulphur dioxide, SO_2	mg/MJ	650	500	1	33	110
Nitrogen oxides, NOx	mg/MJ	500	240	350	50	130
Volatile organic compounds, VOC	mg/MJ	15	7	15	1200	-

Polycyclic aromatic Hydrocarbons, PAH	mg/MJ	2.0	0.25	0.13	5000	25
Dioxins, PCDD, PCDF	mg/MJ	0.04	-	0	-	0.3
Arsenic, As	mg/MJ	8	2	0.0	5	-
Cadmium, Cd	mg/MJ	0.9	0.7	0.0	10	-
Chromium, Cr	mg/MJ	4	1.2	0.0	50	-
Mercury, Hg	mg/MJ	1.7	0.1	0.0	1	0.0
Nickel, Ni	mg/MJ	6	400	0.0	-	-
Lead, Pb	mg/MJ	7.6	25	0.0	200	1.0
Particles	mg/MJ	25	1.3	0	45	20
Resource consumption	g/MJ	39	25	23	100	67

The environmental profile shows the potential contribution to all types of environmental effects that are considered important and undesirable. For each individual release listed in the inventory, its potential contributions are calculated and added to the overall contribution to the relevant effect type. Releases that do not contribute to any known effect type will not influence the environmental profile. The list is obviously subject to change due to developments in scientific knowledge, and, to a certain extent, political initiatives. The list should, therefore, be continually adjusted to accommodate such changes. The table gives a list of the environmental effects that ought to be considered in an assessment:

LCA: Environmental Effect Types

Global	Global warming
	Stratospheric ozone depletion
Regional	Acidification Nutrient enrichment
	Photochemical ozone formation (smog)
	Chronic toxicity
Local	Acute toxicity
	Area degradation, including: risks
	noise smell
	Physical disturbances,
	including:
	soil erosion
	deterioration in landscape quality
	disturbance or destruction of habitats, and hence, ecosystems.

Let us now work out Learning Activity 6.3, before we proceed any further.

Evaluating Environmental Profiles

Evaluation involves a comparison of the alternatives studied (e.g., products, materials and projects), represented by the environmental, resource and the working environment profiles. The objective of the evaluation is to decide which alternative causes the least environmental damage from a life cycle point of view.

If one alternative is better in all the aspects considered, the evaluation is straightforward. In fact, this is often the case when a product is reviewed first with the purpose of reducing its environmental impacts. However, the evaluation situation gets complex when one alternative is better than the rest in some aspects, while another alternative is better in other aspects. This creates the need to give priorities to the different aspects of the loading profiles. For example, we may ask the questions:

- Which is more important – improving performance in the external environment or in the working environment?

- Is reducing the life cycle contribution to ozone depletion more important than diminishing the release of chemicals causing chronic toxicity?

Clearly, no objective answer can be given to questions like these that involve judgement of fundamentally incomparable items. The setting of the priorities involved in the evaluation step inevitably introduces subjective decisions into the LCA. It should be kept in mind that all the work conducted in LCA-steps leading up to the evaluation must be based on objective and scientific principles. The subjective decisions are hence confined to this last step.

When the inventory is complete and all the emissions listed are classified into environmental impact classes, and converted into common effect units, where possible, the result is an environmental profile for each of the alternatives are examined.

Effect Type							Unit
Global							
Greenhouse							CO_2 equivalent
Ozone layer							CFC11 equivalent
Regional							
Acidification							SO_2 equivalent
Nutrient Enrichment							NO_3 equivalent
Toxicity							TOX equivalent
Local							
Acute Toxicity							TOX equivalent
Odour							Smell equivalent
Noise							Square m.
Disrupted Area							Square m.

Comparing the profiles, you may find that Alternative 1 contributes less than Alternative 2 to global warming, acidification and nutrient enrichment, while Alternative 2 contributes less to ozone depletion and acute toxicity.

It may be very difficult to weigh between different classes of environmental impacts. But, it will help, if you know which of the differences in contribution to the environmental effects are important and which are not. An example of the kind of question you may have to answer is: How does 15 kg of CO_2 equivalents compare to 3 kg of CFC – 11 equivalents, or to 5 kg of SO_2 equivalents?

To help in the comparisons of this nature, each of the contribution in the environmental profile may be divided by the emission from a suitable common emission scenario. This scenario may, for instance, be the environmental profile of the total emissions of your state or country. In this case, the resulting normalised environmental profile shown indicates the size of the fraction of national emissions due to each of the environmental effect contributions:

Effect Type		Unit
Global		
Greenhouse		10^{-9}
Ozone layer		10^{-9}
Regional		
Acidification		10^{-9}
Nutrient Enrichment		10^{-9}
Toxicity		10^{-9}
Local		
Acute Toxicity		10^{-9}
Odour		10^{-9}
Noise		10^{-9}
Disrupted Area		10^{-9}

The normalisation step is of particular relevance for comparing alternatives with complicated environmental profiles, i.e., the many different emissions that contribute to different impacts on the environment.

The last step of the LCA is the comparison of the normalised environmental profiles of the different alternatives. The outcome of the evaluation should be the selection of the alternative that performs best from an environmental protection viewpoint.

Different Applications of LCA

Applications for LCA can be broadly classified into private sector applications and governmental applications.

Private Sector Applications

The use of LCA in the private sector varies greatly. To a large extent, this differentiation depends on where a given company is situated in the product chain and the key driver for the LCA activity, e.g., legislation or market competition. For business teams, the LCA tool should be used to understand the environmental issues associated with upstream and downstream processes as well as on-site processes.

Product Development

Using LCA in product development is an obvious choice, as a large part of the future environmental impacts of a product (system) is determined by the design and construction phase. By incorporating LCA in the design phase, companies have the possibility of avoiding or minimising foreseeable impacts compromising the overall quality of the product.

Product development may follow different concepts and routes. The following figure shows some of the common phases of most development methodologies:

Relationships between the Designer's Degree of Freedom and the Level of Information

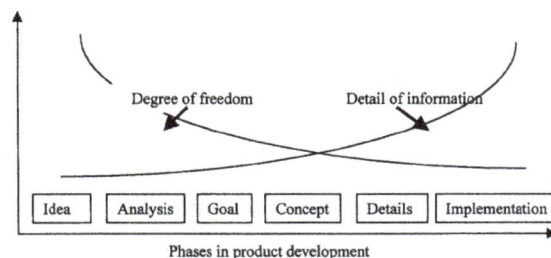

Adapted from Hanssen, 1995.

In the idea-phase, there is almost an unlimited number of possibilities with respect to design, choice of materials, function, etc. The number of options decreases with the development process and changes to the final product and the necessary productive tools often require a whole new development process. It is, therefore, necessary that relevant environmental tools are available and used as early as possible in the development process. For simple products, e.g., packaging, it is possible to apply a detailed and quantitative LCA, since the information on most of the commonly used materials is now available.

Marketing

Marketing is the traditional way of communicating product properties and capabilities, which are consistent with the consumer"s expectations and demands. As the level of environmental consciousness is increasing, the consumer pays more attention to the environmental properties of goods and services. There are four different kinds of environmental marketing. We will describe these, next.

Environmental Labelling (ISO Type-I Labelling)

An environmental label (eco-label) can be seen as a seal of approval for environmentally benign products and can, therefore, be attractive for marketing purposes. Eco-label, at the same time, conveys information to the consumer in a simple but not in an objective way, enabling individuals to include environmental concerns in their own decisions along with considerations for economy and quality.

The general objective of national and supranational eco-labelling scheme is to make products with less environmental impacts visible to the consumer. The success of an eco-labelling scheme, to a certain extent, is dependent on the number of product groups with an eco-label.

Environmental Claims (ISO Type-II Labelling)

An environmental claim is presently defined by ISO (ISO/DIS 14021 and ISO/CD 14022) as a "label or declaration that indicates the environmental aspects of a product or service that may take the form of statements, symbols or graphics on product or packaging labels, product literature, technical bulletins, advertising, publicity or similar applications".

Environmental claims are often uni-dimensional and related to environmental issue of the year or month (e.g., lead-free gasoline, phosphate-free detergent, CFC free hair spray, etc.) Very few of such environmental claims are based on an LCA and, in many cases, focus on irrelevant issues, while failing to address those issues, which are important in a life cycle perspective.

Environmental Declarations (ISO Type-III Labelling)

Environmental declarations may be a tool in eco-marketing to transfer the results from a life cycle investigation of a product (either as a life cycle inventory or a life cycle assessment) to the individual decision-making process of a consumer. The general idea is to give a graphic presentation of a pre-set number of environmental impacts, e.g., by using a bar diagram. Two or more environmental declarations can, on the other hand, illustrate the actual difference between the products and, for environmentally conscious consumers, this can be a valuable supplement to choose between labelled products.

The use of LCA is, thus, a prerequisite for environmental declarations. Standardisation efforts have been initiated by ISO and includes requirements on methodology, transparency, external review, comparative assertions, labelling components, administrative guidelines and procedures governing the accreditation and conduct of type III labelling practitioners.

Organisational Marketing

The classical marketing of environmental performance has mainly been oriented towards products. However, with the increasing number of companies being certified according to ISO 14001, EMAS or BS 7750, some marketing initiatives are being directed towards the environmental capabilities of the companies per se.

As organisations implement the necessary policies for certification, they also encourage formalising the implementation of LCA procedures and life cycle.

Strategic Business Planning

Integration of environmental aspects in strategic business planning is becoming a common feature in many companies. The handling of environmental management schemes such as EMAS (Environmental Management and Auditing Scheme) and ISO 14001 standards. But, many companies still handle the issues on a case-to-case basis.

There are several motivating factors behind the decision to integrate environmental issues, many of which are interrelated. These include, for example, consumer demands, compliance with legislation, community needs for environmental improvement, security of supply and product and market opportunities.

The environmental performance is, thus, changing from being a mandatory property of many products (i.e., all regulatory requirements shall be fulfilled) to being a strong positioning property on the market. An LCA can be used both in relation to existing products and to identify market segments to be opened for environmentally benign products. The LCA information can provide decision-makers with an understanding of the environmental pros and cons of their products and services.

Governmental Applications

Sustainable development has been included as a major item on most of the governmental agendas since the 1992 Rio summit. Although a precise definition of sustainable development has not been given, it is obvious that LCA must be used to ensure that actions towards a more sustainable future will have the desired effect. An LCA, as a specific tool, can ensure this in some cases, while LCA, as an approach or as a strategic tool, can give directions but not the whole answer and must, therefore, be applied along with other tools such as risk assessment, environmental impact assessment, cost benefit analysis and others. The main governmental applications include product oriented policy, deposit-refund schemes, including waste management policies, subsidies and taxation, and general (process-oriented) policies.

Global Warming

This is caused by gases that absorb infra-red radiation (heat) irradiated from the Earth. Increasing their concentrations in the atmosphere may cause the global atmospheric temperature to increase, and this may, through increased melting of polar ice caps and heat-expansion of water in the oceans, lead to rising sea levels and serious climatic changes. The range and the ecological and socio-economic consequences of these effects are impossible to predict at present.

Carbon dioxide, CO_2, is the most prominent of the global warming gases, contributing about 60% of the total anthropogenic (human- made) atmospheric content of global warming gases. Other important gases are methane, CH_4, contributing 15% (the main anthropogenic sources are stocks of cattle and cultivation of rice), dinitrogenoxide (or nitrous oxide), N_2O, with 4% (combustion processes, fertilised agricultural soils) and the halocarbons group, with around 10% (CFCs, HCFCs and halons, all anthropogenic in origin and manufactured for use as solvents, aerosol propellants, heat transmitters, insulators and fire extinguishers).

For carbon dioxide, it is important in the inventory to distinguish between activities that cause a net increase in the atmospheric content (such as combustion of the fossil fuels, oil, coal and gas, or their derivatives such as plastics) and other activities where the CO, CO_2 contribution would reach the atmosphere anyway, through the biogeochemical carbon cycle. The latter includes the burning of biological materials that would otherwise have decayed naturally and released their carbon content as CO_2 to the atmosphere. However, the clearance and burning of forests to create agricultural land is an exception from this rule, since the

activity causes a net reduction in the global biomass and thus a net increase in the atmospheric content of CO_2.

Stratospheric Ozone Depletion

The stratosphere is one of the higher strata of the atmosphere (10 - 50 km above the surface of the Earth). It is characterised by the presence of large concentrations of the gas ozone OCO_2 that constantly exchanges with oxygen through the action of sunlight and various compounds in the stratosphere. The body of ozone in the stratosphere is commonly referred to as the ozone layer. It is of crucial importance for the presence of life as we know it on Earth, since it absorbs most of the ultra-violet (UV) radiations emitted from the sun, which can have very damaging effects, such as causing skin cancers.

It has been found that recent emissions of the man-made halo carbons (described in the paragraph on global warming) are contributing to the breakdown of ozone molecules in the stratosphere, thus favoring ozone degradation over ozone formation and reducing the extent and UV- absorbing capacity of the ozone layer. This may have serious direct effects on life on Earth, especially on the plants and algae that are the primary producers and basis of all food chains. The possible indirect effects are at present hard to assess.

Acidification

Emissions of compounds that contribute to a lowering of the pH in some compartment of the environment are aggregated in this effect type. Acidification mainly occurs in areas where the underground strata are poor in limestone that may neutralise the incoming acids. The biological systems that are affected by acidification are lakes, forests and agriculture. The lowering of the pH in lake water or soil water dramatically changes the living conditions, resulting in massive death of the organisms normally living there (fish, trees, crop plants). Acidification also causes severe economic losses in urban areas, where corrosion of machinery and buildings may be substantially increased.

The contributors to acidification are compounds that are acids themselves (capable of donating protons in chemical reactions) or ones that cause liberations of protons as they experience conversions in a receiving body of the environment. Among the important sources of acidification is combustion of chlorine – containing, fossil fuels, production of metal from sulphide ore, combustion of chlorine substances, e.g., in waste incineration.

Nutrient Enrichment

Nutrient enrichment is the process of enrichment of biotopes (water bodies and soils) in the two main nutrients, nitrogen, N. and phosphorus, P, one of which is generally the nutrient factor limiting biological growth of the system. The effect type incorporates the ensuing changes in chemical and physical living conditions for the organisms and consecutive or cumulative disturbances of the ecosystems. The main focus is on the eutrophication of lakes and coastal waters. Here, the input of nutrients causes algal blooms that shadow out the plant life at the bottom of the water body, even in shallow waters, leading to a vast precipitation of dead algae. The degradation of this additional input of biological material uses up the oxygen dissolved in the water, causing anoxic conditions in the sediments and in the bottom strata. The effect on the fish life and the aquatic ecosystems may lead to dramatic reduction in catches of fish.

Nutrient enrichment of soil ecosystems may cause changes in the species composition of plant communities growing on soils poor in nutrients, thus threatening to extinguish certain plant species.

Through percolation, the nutrient-enriched precipitation may reach the groundwater and cause unwanted concentrations of compounds, particularly of nitrogen, in this source of drinking water.

Contributing to nutrient enrichment are all the compounds containing nitrogen or phosphorus that may be released during their degradation in the environment. Important sources of N and P are agricultural fertilisers and emissions of untreated sewage from urban districts. Combustion processes may also be a significant source of nitrogen.

Photochemical Ozone Formation

Although neurotoxin in themselves, most hydrocarbons will rarely contribute to toxicity in the environment, because their atmospheric half-lives are too short for large concentrations to build up in the troposphere (0 - 10 km above the surface of the Earth). The atmospheric degradation of the hydrocarbons takes place through reactions with nitrogen oxides and hydroxyl radicals present in the troposphere. The reaction schemes include photochemical reactions requiring sunlight, and they lead to the formation of aggressive gaseous compounds, the most important of which is ozone. While the presence of ozone in the higher strata of the atmosphere is indispensable for the existence of life, as we know it, it is highly noxious in the air that we breathe.

Episodes of high concentrations of these reactive gases may occur locally due to emissions of excessive amounts of hydrocarbons, e.g., from traffic during rush hours, but their extent may increase to a regional scale in periods where meteorological conditions prevent the vertical mixing of air within the troposphere. Such episodes are referred to as photo (or photo-chemical) smog.

Ozone is a very reactive molecule, and it will oxidise any reduced compound it encounters including most of bimolecular that are present in the troposphere. Plants in particular suffer from exposure to high ozone concentrations, as they are unable to move during episodes of photo smog, and they depend completely on the exchange of gases with the surrounding air to furnish carbon dioxide for photosynthesis.

Toxicity (Chronic or Acute)

This type of environmental effect includes contributions to toxicity (or poisoning) affecting any kind of living organism or ecosystem (human toxicity as well as environmental toxicity). The toxicity may be acute or chronic, and literally thousands of different chemical compounds may contribute, with toxic effects of many diverse types. Potential contributions to acute or chronic toxicity in the environment are much more difficult to quantify and to aggregate than contributions to any of the other effect types.

Since many different compounds contribute to this environmental effect type, most human activities that might be considered an object for an environmental assessment entail releases that will be toxic in the environment. Therefore, it is difficult to point out individual human activities as the main sources of environmental toxicity. Nevertheless, chemical industries of all types generally cause significant emissions of many different toxins affecting humans and the environment.

Though a quantitative approach to the aggregation or accumulation of toxicity is difficult, this item should under no circumstances be left out of an environmental assessment. Since toxic effects will generally occur on a local scale, the releases contributing to toxicity will often be the ones that are felt most strongly by people living in the surroundings of the project considered. They will, therefore, attract the most attention. Neglecting toxic effects in the classification may strongly reduce the general credibility of the environmental assessment.

Resource Consumption

Apart from reducing the environmental load of the product, a general goal of environmental management of products is to reduce the use of limited or non-renewable resources, so as not to impair the possibilities for future generations on planet Earth. (Current practices should, as a general rule be sustainable.) Therefore, a distinction should be made between renewable and non-renewable resources in the inventory. A renewable resource is defined as one that is being replenished naturally, whereas non- renewable resources are not regenerated at a rate that is significant in a historical perspective.

For historical reasons, use of resources in energy production is considered as a special kind of resource consumption with its own inventory. Also, an inventory of the consumption of energy resources must distinguish between renewable and non- renewable resources.

The bare classification into renewable and non-renewable resources is insufficient for assessment purposes. As a minimum, some further raking of the different kinds of non-renewable resources is necessary. For example, although both sand and copper ore are non-renewable resources, their use should not be considered as equally critical in an LCA. Unfortunately, there is no generally accepted scheme for classification of resources consumption, similar to the scheme adopted for environmental releases.

Life Cycle Analysis Case Study: Substitutes for PVC

After a long, fierce political debate on the desirability and possibility of reducing the consumption of the plastic material PVC, (Poly Vinyl Chloride), The Danish Agency for Environmental Protection commissioned a life cycle assessment of PVC and alternative materials to determine the environmental and health related consequences of substituting other materials for PVC in packaging materials, building materials and office and hospital articles.

The life cycle assessment (LCA) was performed as a screening involving all life cycle stage of the materials, from the extraction of raw materials to their disposal through landfilling, incineration or possible recycling. A more detailed study was performed on what seemed, from the screening, to be the key issues for the life cycle of the material.

This investigation differed from most other LCA studies in assigning equal importance to working environment/health/risk aspects and the external environment.

The system studied was specified by the prevalent conditions in Denmark concerning working environment and safety regulations. The disposal method was considered to be incineration with energy recovery for heating.

The following aspects were studied for each of the materials:

- consumption of resources (including energy),

- potential exposure in the working environment,

- potential effects in the working environment,

- potential exposure in the natural environment,

- potential effects in the natural environment, and risk of accidents.

The materials evaluated were:

- polyvinyl chloride, PVC,

- polyethylene/polypropylene, PE/PP

- polystyrene, PS,

- polyurethane, PUR,

- polyethylene terephthalate, PET,

- vacuum-treated and surface-treated wood,

- paper/cardboard, and

- aluminium.

Several of the investigated materials were only considered as alternatives to PVS in some application. Some conclusions of the study are given below.

Polyvinyl Chloride

- The worst problems related to environmental aspects seemed to be:

- emissions of dioxins and chlorinated organic compounds in the waste water from production and combustion gases from incineration of PVC;

- emission of hydrochloric acid and toxic heavy metals (particularly lead and cadmium) that may be added as stabilisers to PVC for some applications.

Those related to health aspects were:

- exposure to carcinogenic vinyl chloride monomer, which may occur during production of the polymer and its manufacture into products;

- exposure to hydrochloric acid and to the carcinogen plasticiser, DEHP, occurring during manufacture of PVC products;

- risk of exposure to chlorine and carcinogenic vinyl chloride monomer in the event of accidents at PVC polymer production facilities.

Polyethylene/Polypropylene

Unlike PVC, these plastics do not contain chlorine, and this leads to substantial improvements in environmental and working environment performance. The energy consumption for production of PE and PP is similar to that required for PVC. The main problem with these materials is that they require addition of flame retardants from some applications, which may introduce environmental problems in the production, manufacture and disposal of the polymers.

The overall conclusion, however, is that the substitution of PVC with polyethylene or polypropylene, where possible, will give substantial improvements in external and working environmental performance.

Polystyrene

The production of polystyrene involves a risk of exposure to the reproduction-damaging substance, styrene. The polymerisation of styrene to polystyrene may involve a risk of explosion. The energy consumption for production of polystyrene is considerably higher than for PVC.

Polystyrene may be expanded into a rigid foam by blowing with gases. In some cases, the gases used are the highly ozone- degrading CFC or HCFC gases. If use of CFCs and HCFCs is avoided, the substitution of polystyrene for PVC will lead to environmental improvement, particularly at the disposal stage. The working environment aspects are unclear.

Polyethylene Terephthalate

PET entails somewhat higher energy consumption than PVC. The overall environmental performance is clearly better for PET than for PVC, both for working environment, external environment and probability and consequences of accidents.

Polyurethane

The production of PUR involves the use of isocyanides, exposure to which is extremely unpleasant in the working environment during production of the polymer and manufacture into products.

PUR may, similarly to polystyrene, be expanded by blowing with CFCs or HCFCs. If this application is disregarded, the potential effects of PUR in the external environment seem smaller than for PVC, particularly at the disposal stage (due to the absence of chlorine and toxic metals). For working environment conditions, however, there is some aggravation compared to PVC. Overall, the report does not recommend a substitution of PVC with PUR.

Vacuum-treated and Surface-treated Wood

In contrast to all the polymer materials, wood is a renewable resource and generally, little or no fossil energy is required for its production. The manufacture of vacuum-treated wood, e.g., into window frames, involves the exposure of works to carcinogenic wood dust, and the vacuum-treatment of the wood entails emissions of heavy metals into the external environment.

Nevertheless, the overall conclusion qualifies vacuum- or surface- treated wood as an environmentally advantageous alternative to PVC.

Paper/cardboard

Like wood, paper and cardboard are made from renewable resources. Their energy consumption is higher than for wood, but still lower than for PVC.

The environmental profile of paper and cardboard depends strongly on the process used. Often, the pulp is bleached using chlorine compounds, resulting in emission of chlorinated organic compounds (including dioxins) in the wastewater. Generally, pulp production involves emissions of large quantities of toxic and oxygen-consuming organic material in the wastewater that is worse than the loading from the PVC-production. The disposal of paper or cardboard after its use is substantially better from an environmental point of view, and overall, paper or cardboard is recommended over PVC for packaging purposes and similar applications.

Aluminium

Production of primary aluminium from bauxite is extremely energy- consuming and involves some serious working environment problems and hazards. Together these aspects disqualify the substitution of PVC with primary aluminium.

For secondary aluminium, i.e., aluminium recycled through collection and recasting, the situation is different since it may be recycled, and the loading of the original production of primary aluminium reduced correspondingly. The preference of PVC or aluminium thus depends on the efficiency of the recycling of aluminium.

Life Cycle Analysis: A Case of Steel Sector in India

Since 1999, with an objective of optimum utilisation of resources, Life Cycle Assessment Study in the Steel Sector has been continued with joint funding of the Ministry and three identified steel plants, viz, Bhilai Steel Plant, SAIL; Jamshedpur Steel Plant of TISCO and Vizag Plant of Rashtriya Ispat Nigam Ltd. (RINL). The salient findings of the study are summarised herein.

Scope of the Study

- Cradle to Gate (RM mining to Crude Steel)
- Includes Inventory Analysis
- Impact assessment
- Improvement Analysis

Boundary of the Study

- Crude Steel at Factory Gate
- Quantification Unit - Kg of crude steel at factory gate
- Includes:
- Steel Making
- Production & Transportation of RM

- Production & Transportation of Energy

- Production & Transportation of Consumables

Data Collection & Validation

- Main Source of Data - Annual Statistics

- Coke Oven Gas, BF Gas and BOF Gas conversion from calories to joules by 4.1868 joules/cal

- Heat Content of 8 ata steam 3000 MJ/tonne and 18 ata 3260MJ/tonne

- Fugitive Dust and drinking water not included

- Data Validation for FY1997 & FY 1998 by MECON   Primary data validation by log sheet checks

- Secondary data validation - Monthly, Quarterly, Half Yearly &annual Audits

- Energy Audits - Steam, Electricity, Fuel

- Emission Audits - Weekly Stack & effluent analysis

- Performance Audits - Daily Production Target monitoring at department level and against techno-economic parameters at plant level

- Technology Audits

Results

Sl. No.	Item	World Av. (Kg/Kg of steel)	Units Av. (Kg/Kg of steel)
1.	Coal	0.536	0.989
2.	Dolomite	0.0291	0.158
3.	Limestone	0.0001	0.171
4.	Iron Ore	1.38	1.813
5.	Steel Scrap	0.142	0.110
6.	Water	8.1(l/kg)	4.74 (l/kg)
7.	Energy	21.7 (MJ/kg)	29.77 (MJ/Kg)

Emission Results

Sl. No.	Item	World Av.(g/Kg of steel)	Units Av (g/Kg of steel)
1.	CO_2	1734	2924.65
2.	CO	26.4	63.09
3.	NOx	1.57	1.39
4.	SOx	1.48	1.70
5	Particulates	1.57	1.275

Result Analysis

- Ash content of Domestic coal high so coal consumption high.

- Alumina in Domestic ore high so higher Dolomite requirement.

- Iron Ore lump used in cooling in BOF in India nowhere else in the world so higher Iron Ore Consumption.

- Steel plants outside India mostly on shores, don't recycle water, hence high consumption.

- World over steel plants use BF gas top pressure to extract electricity not in this unit.

- BF gas having higher Carbon though lower calorific value compared to other gases used so higher CO_2 emission.

- World over alternate fuels used in BF like oil, Natural Gas, Pulverised Coal etc. BF gas generated per tonne is low.

- High CO emission in Sinter making process.

- Low Sulphur in Indian Coals so low SOx emission.

- ODS use is low.

Impact Assessment

Global Warming effect (direct, 20 years)

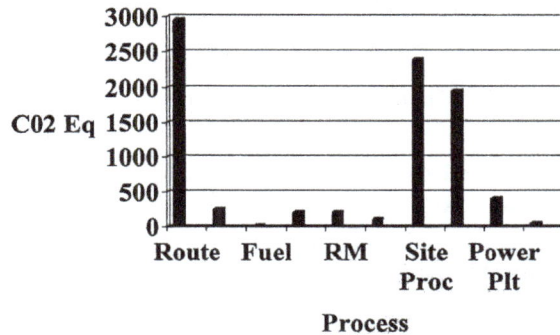

Air Acidification Effect

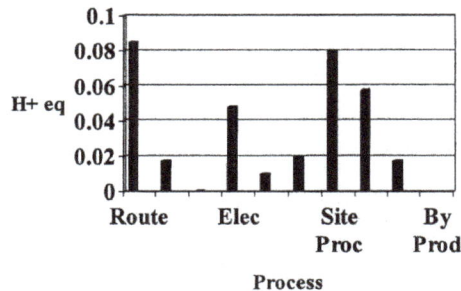

Improvement Planning

- Increase in proportion of imported coal in coal blend

- New Sinter machine installation

- Installation of Top Gas Pressure Recovery to harness electricity

- Coal Dust Injection in Blast Furnace

- Increased Recovery of BOF gas

References

- Curkovic, S, and Sroufe, R.P., "Using ISO 14001 to Promote a Sustainable Supply Chain Strategy," accepted, International Journal of Business Strategy and the Environment, Vol. 20, No. 2, 71-93, 2011

- "ISO 14000 family - Environmental management". ISO. International Organization for Standardization. Retrieved 22 May 2017

- Hendrickson, C. T., Lave, L. B., and Matthews, H. S. (2005). Environmental Life Cycle Assessment of Goods and Services: An Input–Output Approach, Resources for the Future Press ISBN 1-933115-24-6

- Finnveden, G., Hauschild, M.Z., Ekvall, T., Guinée, J., Heijungs, R., Hellweq, S., Koehler, A., Pennington, D. & Suh, S. (2009). Recent developments in Life Cycle Assessment. Journal of Environmental Management 91(1), 1-21

- "ISO 45001 - Occupational health and safety". ISO.org. International Organisation for Standardization. Retrieved 2016-11-18

- Jiménez-González, C.; Kim, S.; Overcash, M. Methodology for developing gate-to-gate Life cycle inventory information. The International Journal of Life Cycle Assessment 2000, 5, 153–159

- "PAS 2050:2011 Specification for the assessment of the life cycle greenhouse gas emissions of goods and services". BSI. Retrieved on: 25 April 2013

- S. Singh; B. R. Bakshi (2009). "Eco-LCA: A Tool for Quantifying the Role of Ecological Resources in LCA". International Symposium on Sustainable Systems and Technology: 1–6. doi:10.1109/ISSST.2009.5156770. ISBN 978-1-4244-4324-6

- Curran, Mary Ann. "Life Cycle Analysis: Principles and Practice" (PDF). Scientific Applications International Corporation. Archived from the original (PDF) on 18 October 2011. Retrieved 24 October 2011

- Sciubba, E (2004). "From Engineering Economics to Extended Exergy Accounting: A Possible Path from Monetary to Resource-Based Costing" (PDF). Journal of Industrial Ecology. 8 (4): 19–40. doi:10.1162/1088198043630397

- Brinkman, Norman; Eberle, Ulrich; Formanski, Volker; Grebe, Uwe-Dieter; Matthe, Roland (15 April 2012). "Vehicle Electrification - Quo Vadis". VDI. Retrieved 27 April 2013

Permissions

Index

www.ingramcontent.com/pod-product-compliance
Lightning Source LLC
Chambersburg PA
CBHW061245190326

41458CB00011B/3586